Biodiversity of Vegetation and Flora in Tropical Africa

Biodiversity of Vegetation and Flora in Tropical Africa

Editors

Luís Catarino
Maria M. Romeiras

MDPI • Basel • Beijing • Wuhan • Barcelona • Belgrade • Manchester • Tokyo • Cluj • Tianjin

Editors
Luís Catarino
University of Lisbon
Portugal

Maria M. Romeiras
University of Lisbon
Portugal

Editorial Office
MDPI
St. Alban-Anlage 66
4052 Basel, Switzerland

This is a reprint of articles from the Special Issue published online in the open access journal *Diversity* (ISSN 1424-2818) (available at: https://www.mdpi.com/journal/diversity/special_issues/biodiversity_vegetation_flora).

For citation purposes, cite each article independently as indicated on the article page online and as indicated below:

LastName, A.A.; LastName, B.B.; LastName, C.C. Article Title. *Journal Name* **Year**, *Article Number*, Page Range.

ISBN 978-3-03943-531-9 (Hbk)
ISBN 978-3-03943-532-6 (PDF)

Cover image courtesy of Luís Catarino.

Contents

About the Editors . . . . **vii**

Luís Catarino and Maria M. Romeiras
Biodiversity of Vegetation and Flora in Tropical Africa
Reprinted from: *Diversity* **2020**, *12*, 369, doi:10.3390/d12100369 . . . . **1**

Iveren Abiem, Gabriel Arellano, David Kenfack and Hazel Chapman
Afromontane Forest Diversity and the Role of Grassland-Forest Transition in Tree
Species Distribution
Reprinted from: *Diversity* **2020**, *12*, 30, doi:10.3390/d12010030 . . . . **5**

**Carlos Neto, José Carlos Costa, Albano Figueiredo, Jorge Capelo, Isildo Gomes,
Sónia Vitória, José Maria Semedo, António Lopes, Herculano Dinis, Ezequiel Correia,
Maria Cristina Duarte and Maria M. Romeiras**
The Role of Climate and Topography in Shaping the Diversity of Plant Communities in Cabo
Verde Islands
Reprinted from: *Diversity* **2020**, *12*, 80, doi:10.3390/d12020080 . . . . **25**

**Motiki M. Mofokeng, Hintsa T. Araya, Stephen O. Amoo, David Sehlola,
Christian P. du Plooy, Michael W. Bairu, Sonja Venter and Phatu W. Mashela**
Diversity and Conservation through Cultivation of *Hypoxis* in Africa—A Case Study of
Hypoxis hemerocallidea
Reprinted from: *Diversity* **2020**, *12*, 122, doi:10.3390/d12040122 . . . . **41**

**John L. Godlee, Francisco Maiato Goncalves, José João Tchamba, Antonio Valter Chisingui,
Jonathan Ilunga Muledi, Mylor Ngoy Shutcha, Casey M. Ryan, Thom K. Brade and
Kyle G. Dexter**
Diversity and Structure of an Arid Woodland in Southwest Angola, with Comparison to the
Wider Miombo Ecoregion
Reprinted from: *Diversity* **2020**, *12*, 140, doi:10.3390/d12040140 . . . . **65**

**Baptiste Brée, Andrew J. Helmstetter, Kévin Bethune, Jean-Paul Ghogue, Bonaventure Sonké
and Thomas L. P. Couvreur**
Diversification of African Rainforest Restricted Clades: Piptostigmateae and Annickieae
(Annonaceae)
Reprinted from: *Diversity* **2020**, *12*, 227, doi:10.3390/d12060227 . . . . **85**

**Iain Darbyshire, Carrie A. Kiel, Corine M. Astroth, Kyle G. Dexter, Frances M. Chase and
Erin A. Tripp**
Phylogenomic Study of *Monechma* Reveals Two Divergent Plant Lineages of Ecological
Importance in the African Savanna and Succulent Biomes
Reprinted from: *Diversity* **2020**, *12*, 237, doi:10.3390/d12060237 . . . . **103**

Maame Esi Hammond and Radek Pokorný
Diversity of Tree Species in Gap Regeneration under Tropical Moist Semi-Deciduous Forest:
An Example from Bia Tano Forest Reserve
Reprinted from: *Diversity* **2020**, *12*, 301, doi:10.3390/d12080301 . . . . **129**

Silvia Catarino, Maria Manuel Romeiras, Rui Figueira, Valentine Aubard, João M. N. Silva and José M. C. Pereira
Spatial and Temporal Trends of Burnt Area in Angola: Implications for Natural Vegetation and Protected Area Management
Reprinted from: *Diversity* **2020**, *12*, 307, doi:10.3390/d12080307 . . . . **149**

About the Editors

Luís Catarino, Ph.D., is a researcher in the Plant Biology Department in the Faculty of Sciences of Lisbon University and member of the Centre for Ecology, Evolution and Environmental Changes (cE3c). His main research interests are tropical vegetation ecology, flora and land cover changes; ethnobotany, ethnoecology and sustainable use of natural resources; and protected area management and monitoring. He has led and participated in several research projects and consultancies in Tropical Africa (Angola, Guinea-Bissau, Cape Verde) and Tropical South America (Brazilian Legal Amazon), resulting in the publication of scientific articles, book chapters and books, as well as the presentation of communications at scientific meetings. He is the supervisor of master's and doctoral theses on vegetation ecology, ethnobotany, flora and natural resource assessment in Guinea-Bissau and Angola.

Maria M. Romeiras, Ph.D., is Professor at University of Lisbon's Instituto Superior de Agronomia (ISA), the largest and most qualified school of graduate and postgraduate degrees in the agricultural sciences in Portugal. Currently, she is member of two research centers: full member in "Linking Landscape, Environment, Agriculture and Food (LEAF/ISA)" and collaborator in "Centre for Ecology, Evolution and Environmental Changes (cE3c/FCUL)". Maria has published over 12 book chapters and more than 60 papers in international journals. Currently, she is supervising several graduate students (5 Ph.D. students and 15 M.S. students). Over the last years, she organized several field expeditions, particularly several in Cabo Verde and Angola. Maria has participated in several research projects and recently was Principal Investigator in 2 projects focused on the study of African flora. Maria's research interests are mainly focused on tropical plant diversity, particularly in molecular systematics, conservation, island biogeography, agrobiodiversity and ethnobotany.

Editorial

Biodiversity of Vegetation and Flora in Tropical Africa

Luís Catarino [1,*] **and Maria M. Romeiras** [1,2,*]

1   Centre for Ecology, Evolution and Environmental Changes (cE3c), Faculty of Sciences (FCUL),
    University of Lisbon, 1749-016 Lisbon, Portugal
2   Linking Landscape, Environment, Agriculture and Food (LEAF), School of Agriculture (ISA),
    University of Lisbon, 1349-017 Lisbon, Portugal
*   Correspondence: lmcatarino@fc.ul.pt (L.C.); mmromeiras@isa.ulisboa.pt (M.M.R.)

Received: 24 September 2020; Accepted: 24 September 2020; Published: 25 September 2020

Abstract: African ecosystems comprise a wealthy repository of biodiversity with a high proportion of native and endemic plant species, which makes them biologically unique and providers of a wide range of ecosystem services. A large part of African populations, in both rural and urban areas, depends on plants for their survival and welfare, but many ecosystems are being degraded, mostly due to the growing impacts of climate change and other anthropogenic actions and environmental problems. Loss of habitat and biodiversity affects livelihoods, water supply and food security, and reduces the resilience of ecosystems in the African continent. Knowledge of the huge African plant and ecosystem diversity, and on the structure, composition and processes involved in vegetation dynamics, is crucial to promote their sustainable use and to preserve one of the most understudied regions in the world. This Special Issue aimed to gather contributions that update and improve such knowledge.

Keywords: flora; vegetation; Africa; tropical biodiversity

African biodiversity has been declining over the last decades, with losses of species and habitats. According to UNEP-WCMC [1], ongoing loss of biodiversity in Africa is driven by a combination of human-induced factors that result in deforestation and forest degradation, entailing important land-cover changes. Moreover, the negative impacts of climate change on species and ecosystems are exacerbating the effects of such pressures. Different figures have been published, but the most optimistic calculations indicate a forest loss of 22% across tropical Africa since 1900, whereas the most pessimistic estimates point to 35–55%, to which the large-scale forest degradation must be added [2]. Consequently, as much as one-third of the tropical African flora is potentially threatened with extinction [3].

The contributions to this Special Issue of *Diversity* deal with a range of subjects on African plant and vegetation biology and ecology, both in continental and insular regions. Two main broad subjects are addressed: structure, ecology and composition of African vegetation, and species diversity and conservation.

Among those on the first broad topic, Abiem et al. [4] present a study on forest diversity and distribution of tree species in a grassland–forest transition, a typical ecotone across the continent. The diversity and structure of *miombo*, one of the most important ecoregions in the continent, is studied in southwest Angola by Godlee et al. [5] and compared to other *miombo* sub-types. In a work concerning the Cabo Verde Islands (West Africa), Neto et al. [6] discuss the role of abiotic factors, particularly climate and topography, in shaping the diversity of plant communities in this semi-arid archipelago. Hammond and Pokorný [7] address the process of succession in a tropical moist semi-deciduous forest, analyzing the diversity of tree species in gap regeneration. The important issue of fire effect on natural vegetation and its implications is presented for Angola by Catarino et al. [8], who analyzed the spatial and temporal trends of burnt areas and found increasing trends associated with savannas and grasslands in *miombo* woodlands' biome.

Aspects of species diversity and conservation are addressed, among others, by Brée et al. [9], who analyzed the process of species diversification in African rainforests for two restricted clades of Annonaceae, an important family in African forests. The subject of conservation through cultivation of widely used wild species was examined by Mofokeng et al. [10] with a case study on *Hypoxis hemerocallidea*. With an innovative perspective, a phylogenomic study of the Acanthaceae genus *Monechma*, which occurs in savanna and succulent biomes, is also included. In this study, Darbyshire et al. [11] found divergent plant lineages with ecological importance, and set clues for future research towards understanding the biogeographical history of continental Africa.

The guest editors of this Special Issue of Diversity journal hope that the set of works here gathered will contribute to enhancing knowledge and improving conservation actions concerning the flora and vegetation of tropical Africa.

Funding: The authors are grateful to the Foundation for Science and Technology (FCT, Portugal) for financial support through national funds FCT/MCTES to Centre for Ecology, Evolution and Environmental Changes (cE3c: UIDB/00239/2020), and Linking Landscape, Environment, Agriculture and Food (LEAF: UID/AGR/04129/2020). Also, M.M.R. would like to acknowledge the support provided by FCT and Aga Khan Development Network (AKDN) under the project CVAgrobiodiversity/333111699.

Conflicts of Interest: The authors declare no conflict of interest.

References

1. UNEP-WCMC. *The State of Biodiversity in Africa: A mid-term review of progress towards the Aichi Biodiversity Targets*; UNEP-WCMC: Cambridge, UK, 2016; ISBN 978-92-807-3508-6.
2. Aleman, J.C.; Jarzyna, M.A.; Staver, A.C. Forest extent and deforestation in tropical Africa since 1900. *Nat. Ecol. Evol.* **2017**, *2*, 26–33. [CrossRef] [PubMed]
3. Stévart, T.; Dauby, G.; Lowry, P.P.; Blach-Overgaard, A.; Droissart, V.; Harris, D.J.; Mackinder, B.A.; Schatz, G.E.; Sonké, B.; Sosef, M.S.M.; et al. A third of the tropical African flora is potentially threatened with extinction. *Sci. Adv.* **2019**, *5*, eaax9444. [CrossRef] [PubMed]
4. Abiem, I.; Arellano, G.; Kenfack, D.; Chapman, H.M. Afromontane Forest Diversity and the Role of Grassland-Forest Transition in Tree Species Distribution. *Diversity* **2020**, *12*, 30. [CrossRef]
5. Godlee, J.L.; Gonçalves, F.M.; Tchamba, J.J.; Chisingui, A.V.; Muledi, J.I.; Shutcha, M.N.; Ryan, C.; Brade, T.K.; Dexter, K.G. Diversity and Structure of an Arid Woodland in Southwest Angola, with Comparison to the Wider Miombo Ecoregion. *Diversity* **2020**, *12*, 140. [CrossRef]
6. Neto, C.; Costa, J.C.; Figueiredo, A.; Capelo, J.H.; Gomes, I.; Silva, S.; Semedo, J.M.; Lopes, A.; Dinis, H.; Correia, E.; et al. The Role of Climate and Topography in Shaping the Diversity of Plant Communities in Cabo Verde Islands. *Diversity* **2020**, *12*, 80. [CrossRef]
7. Hammond, M.E.; Pokorný, R. Diversity of Tree Species in Gap Regeneration under Tropical Moist Semi-Deciduous Forest: An Example from Bia Tano Forest Reserve. *Diversity* **2020**, *12*, 301. [CrossRef]
8. Catarino, S.; Romeiras, M.M.; Figueira, R.; Aubard, V.; Silva, J.; Pereira, J.M.C. Spatial and Temporal Trends of Burnt Area in Angola: Implications for Natural Vegetation and Protected Area Management. *Diversity* **2020**, *12*, 307. [CrossRef]
9. Brée, B.; Helmstetter, A.J.; Bethune, K.; Ghogue, J.-P.; Sonké, B.; Couvreur, T.L. Diversification of African Rainforest Restricted Clades: Piptostigmateae and Annickieae (Annonaceae). *Diversity* **2020**, *12*, 227. [CrossRef]

10. Mofokeng, M.M.; Araya, H.T.; Van Staden, J.; Sehlola, D.; Du Plooy, C.P.; Bairu, M.W.; Venter, S.; Mashela, P.W. Diversity and Conservation through Cultivation of Hypoxis in Africa—A Case Study of Hypoxis hemerocallidea. *Diversity* **2020**, *12*, 122. [CrossRef]
11. Darbyshire, I.; Kiel, C.A.; Astroth, C.M.; Dexter, K.G.; Chase, F.M.; Tripp, E.A. Phylogenomic Study of *Monechma* Reveals Two Divergent Plant Lineages of Ecological Importance in the African Savanna and Succulent Biomes. *Diversity* **2020**, *12*, 237. [CrossRef]

Article

Afromontane Forest Diversity and the Role of Grassland-Forest Transition in Tree Species Distribution

Iveren Abiem [1,2,3,*], **Gabriel Arellano** [4,5], **David Kenfack** [5] **and Hazel Chapman** [1,3]

1 School of Biological Sciences, University of Canterbury, Private Bag 4800, Christchurch 8140, New Zealand; hazel.chapman@canterbury.ac.nz
2 Department of Plant Science and Biotechnology, University of Jos, P.M.B. 2084, Jos 930001, Nigeria
3 Nigerian Montane Forest Project, Yelwa Village 663102, Taraba State, Nigeria
4 Ecology and Evolutionary Biology, University of Michigan, Ann Arbor, MI 48109, USA; gabriel.arellano.torres@gmail.com
5 Forest Global Earth Observatory, Smithsonian Tropical Research Institute, Washington, DC 20036, USA; KenfackD@si.edu
* Correspondence: abiemiveren@gmail.com; Tel.: +234-70-3165-9327

Received: 7 November 2019; Accepted: 7 January 2020; Published: 15 January 2020

Abstract: Local factors can play an important role in defining tree species distributions in species rich tropical forests. To what extent the same applies to relatively small, species poor West African montane forests is unknown. Here, forests survive in a grassland matrix and fire has played a key role in their spatial and temporal dynamics since the Miocene. To what extent these dynamics influence local species distributions, as compared with other environmental variables such as altitude and moisture remain unknown. Here, we use data from the 20.28 ha montane forest plot in Ngel Nyaki Forest Reserve, South-East Nigeria to explore these questions. The plot features a gradient from grassland to core forest, with significant edges. Within the plot, we determined tree stand structure and species diversity and identified all trees ≥1 cm in diameter. We recorded species guild (pioneer vs. shade tolerant), seed size, and dispersal mode. We analyzed and identified to what extent species showed a preference for forest edges/grasslands or core forest. Similarly, we looked for associations with elevation, distance to streams and forest versus grassland. We recorded 41,031 individuals belonging to 105 morphospecies in 87 genera and 47 families. Around 40% of all tree species, and 50% of the abundant species, showed a clear preference for either the edge/grassland habitat or the forest core. However, we found no obvious association between species guild, seed size or dispersal mode, and distance to edge, so what leads to this sorting remains unclear. Few species distributions were influenced by distance to streams or altitude.

Keywords: ecotone; fire; forest core; habitat preference; Ngel Nyaki; niche partitioning; savannah; species sorting; torus translation

---

1. Introduction

Afromontane forests are widespread across the African highlands [1–3]. Typically occurring above 1500 m in elevation, they extend from the Arabian Peninsula south along the rift to the Drakensberg Mountains in the east. In Western Africa, they are restricted to the Cameroon volcanic line and the Guinea highlands [1,4] (Figure 1). This disjunct distribution is ancient, pre-dating the Miocene doming that resulted in the Rift Valleys and the aridification of the East African interior [5]. Despite this distribution, Afromontane forests have a distinctive fauna and flora and harbor a high proportion of endemic species [6–8]. They have substantial carbon stocks [9], provide watershed protection [10] and other important ecosystem services [11]. However, Afromontane forests tend to occur in areas of high human population density and are under intense threat from agricultural practices, fire, and grazing [7,12]. Long-term monitoring of these forests is crucial for understanding how they function and for predicting their dynamics, essential information for sustainable management and preservation of the local and global services they provide.

Afromontane forests are characteristically small and fragmented; they typically occur within a grassland matrix and in West Africa the relative extent of grassland vs. forest has shifted in response to climate fluctuations since the Pleistocene [1,8,13]. Grasslands have dominated over the past 5000 years [14,15], and especially the last 3000 years [16]. Today, a combination of local climate and fire dynamics dictate the balance between grassland and forest [1,17–19], with the forest/grassland boundary reflecting the equilibrium between two processes. First, the progressive woody plant encroachment from the forest into the savannah/grassland, with propagules of woody species taking advantage of high-light conditions and lack of competition in the savannah [20,21]. Second, the elimination of woody plant seedlings, saplings and juveniles by grassland fires. Burning leads to a sharp boundary between forest and savannah/grassland [22,23], which is especially pronounced when fires are lit annually and used as a management tool for cattle grazing. In contrast, if fire is suppressed sufficiently, woody plants colonize the savannah/grassland forming narrow but distinct species-rich transition zones. Eventually, given sufficient time and protection from fire, the woody plant encroachment process continues and forests form [20,24]. Once established, the forest remains relatively self-protected from burning because of the lack of fine grass fuel required for intense/frequent fires, more shady conditions, and in general a different microclimate that makes it less prone to burn [25].

The historic dynamic equilibrium between forest and fire has major implications for forest structure and dynamics [8]. Having survived for thousands of years as islands of forest within a grassland matrix, Afromontane forests have long experienced strong edge effects. How these effects filter species and traits and thus impact forest composition is under investigation in the Afro-temperate forests of the Drakensberg Mountains of South Africa [1,26]. To what extent the South African situation applies in West Africa is currently unknown, although the fact that Le'zine, et al. [10] have recently confirmed extreme climate-related fluctuations in forest cover in the Cameroon highlands (the focus of this study) suggests a similar scenario. To begin to answer this question, we document tree species composition within a 20.28 ha permanent forest plot comprising mid-altitude montane forest and grassland, and explore correlations between tree species distributions and four environmental gradients: the proportion of grass surrounding individual trees (forest versus grassland), proximity to the grassy matrix/forest edge, elevation, and distance to streams (a proxy for soil moisture). In our study we tested the hypothesis that we would find a gradient in tree species composition and functional traits reflecting the core-edge gradient. We thought species sorting from the core to the edge should result from the progressive encroachment of forest into the grassland matrix via dispersal of small seeded, light tolerant species [26].

Finally, because West African montane forests are relatively understudied compared with those elsewhere in Africa, we compare our inventory statistics with those of other Afromontane forests.

2. Materials and Methods

2.1. Study Area and Study Site

The study was conducted in the 260 × 780 m (20.28 hectares) Ngel Nyaki Forest Dynamics Plot, part of the Smithsonian Forest Global Earth Observatory network (Forest-GEO). The plot coordinates are 07°04′05″ N in latitude and 11°03′24″ E in longitude, with an elevation range of 1588 m to 1690 m. The plot is within the 5.7 km^2 Ngel Nyaki forest, located within the 46 km^2 Ngel Nyaki Forest Reserve on the Mambilla Plateau, Taraba State [27,28] (Figure 1). The Nigerian Montane Forest Project (NMFP) [29] is based on the edge of the forest and collects regular weather data. The mean annual rainfall is 1800 mm, with most rain falling between April and October and a six-month dry season. The mean annual temperature is 19 °C and the monthly mean maximum and minimum temperatures for the wet and dry seasons are 25.6 and 15.4 °C, and 28.1 and 15.5 °C, respectively (NMFP weather data). The soil in Ngel Nyaki forest is volcanic and has a high clay content and a pH ranging between 6 and 6.5 [27]. Despite its relatively high nutrient content, the forest soil is easily leached.

Ngel Nyaki forest is floristically diverse and contains six International Union for the Conservation of Nature (IUCN) Red-Listed tree species including *Entandophragma angolense* (Meliaceae), *Lovoa trichiliodes* (Meliaceae), *Eugenia gilgii* (Myrtaceae), *Dombeya ledermannii* (Malvaceae), *Prunus africana* (Rosaceae), and *Khaya grandifoliola* (Meliaceae) [30]. It is a Birdlife International Important Bird Area with several species of large-gaped birds such as hornbills (*Tockus alboterminatus*) and turacos (*Tauraco* spp.). Primate species include the endangered Nigeria-Cameroon chimpanzee (*Pan troglodytes* ssp. *ellioti*), the putty-nosed monkey (*Cercopithecus nictitans*), and the tantalus monkey (*Chlorocebus tantalus*). Smaller mammals include civet cats (*Civettictis civetta*) and rodents such as the African giant pouched rat (*Cricetomys* spp.). While the forest is a State Forest Reserve and therefore protected from hunting and grazing, in practice there is very little protection. To what extent past human disturbance has influenced species distribution is unclear. Forest edges have been farmed on the lower slopes of the forest and cattle have damaged a substantial proportion of the reserve. Within the plot however, there is little evidence of disturbance other than past cattle incursions at the edge.

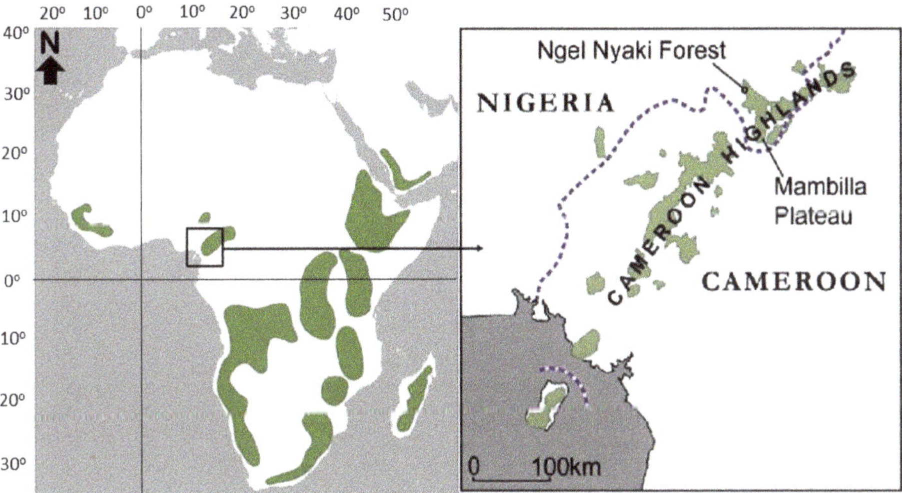

Figure 1. Location of the Ngel Nyaki Forest, on the Mambilla Plateau, Nigeria relative to the approximate distribution of other major montane ecosystems in Africa, shown in green [31]. The forest lies close to the Nigeria/Cameroon border, represented by a dashed purple line.

2.2. Census

The plot was established and measured following the standard protocol of the Smithsonian Center for Tropical Forest Science [32]. We measured, tagged and identified to morphospecies, every free-standing woody plant with diameter at breast height (dbh) ≥1 cm between January 2015 and January 2016. Stem diameter was recorded at 1.3 m above the ground, except when buttressed or swollen. Measurements were taken 50 cm above large buttresses and 2 cm below the lowest point of swellings. Trees were mapped with the highest possible precision in the field, typically at 0.1 m resolution within 5 × 5 m quadrats. Species identification was carried out mostly in the field and supplemented with DNA barcoding. Voucher specimens were collected for each morphospecies and at least one duplicate voucher is held at the NMFP herbarium. DNA barcode sequences of all morphospecies are deposited in the Barcode of Life Data System [33].

2.3. Environmental Variables and Gradients Evaluated

We calculated the location of each individual tree along three "gradients"; grassland edge to core forest, distance to water, and elevation. The elevation was based on detailed topographic maps obtained during the installation of the plot. We also mapped the two streams that run through the plot. Whether a tree was located in "forest vs. grass" was based on a geo-located Google Earth [34] image. The image comprised channels in the RGB space. An area was considered to be "grass" when the red content in a 1 × 1 m pixel in a jpg image was 37% or more. This 37% cutoff in the red channel resulted in a binary habitat map for the area with a very precise concordance with the grass vs. forest distinction on the ground (Figure 2). To avoid spurious results when calculating the minimum distance to grass from each individual tree, we removed isolated grass-looking pixels within the forested area. We forced all pixels with known presence of trees to be "forest" pixels, so all the trees were, by definition, at a distance >0 from the grass.

After defining the environment, we calculated four variables for each individual tree:

(1)   The proportion of grass in a 20 × 20 m area surrounding the individual tree. Rather than count the grass vs. forest pixels in a 20 m diameter circle around each tree, we used a 20 × 20 m square window around each tree because it was computationally more efficient than defining a circle around each tree. Within this square we counted the number of red (grass) vs. other 1 × 1 m pixels and from this calculated the proportion of "grass" vs. forest pixels. We used this metric to indicate whether a species was a forest or grassland species. Our 20 × 20 m squares overlapped and were not necessarily independent of each other.

(2)   The distance to the closest grass pixel. This was calculated from the coordinates of each individual and the grass vs. forest binary map. This metric indicates if a species prefers the forest core (=far from grass) or the forest edge or the grassland itself (=close to the grass).

(3)   The shortest distance to any of two streams in the plot.

(4)   The local elevation.

2.4. Species Guilds

Our five-guild classification is modified from Hawthorne [35,36]: (1) pioneer; (2) non-pioneer light demanding—needs shade to germinate and establish but larger individuals need light; (3) cryptic pioneer—needs light to germinate and establish but can survive for many years as adult in the understory; (4) shade-tolerant; and (5) grassland. For species not included in Hawthorne's 1995 Ghanaian study [35], we identified species guild from information on the species in the JSTOR online plant database [37].

Our seed weight data came from data collected on fruit and seed traits as part of the Nigerian Montane Forest Project [29] ongoing research. Twenty seeds from fruits of at least five individual trees across the forest were weighed to the nearest g and the average seed weight taken. Dispersal modes were based on fruit and/or seed morphology, as described in [38]. Fleshy fruits were considered animal dispersed and hard seeds in pods, ballistically dispersed. Winged seeds are wind dispersed and very heavy, large seeds embedded in a thick husk, gravity dispersed. Growth form, i.e., tree, shrub or strangler, and position in canopy (emergent, canopy, or understory) was based on [27]. In addition, trees growing in the grassland with obvious adaptations to fire were classified as savannah trees.

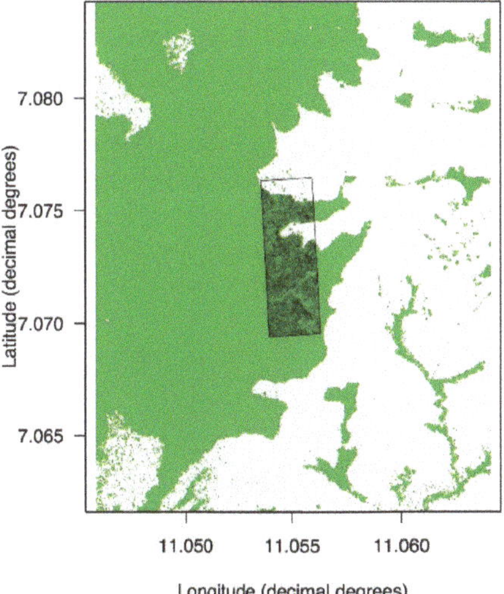

Figure 2. Vegetation mosaic in the Ngel Nyaki Forest Dynamics Plot (260 × 780 m) and surrounding areas. White pixels are grass, green pixels are forest. The location of the plot is shown by the black rectangle on the simplified vegetation map on the left (**A**). For the purposes of this study, "grass" was defined as pixels with red content 37% or more in a jpg image of the area taken from Google Earth, presented on the right (**B**). This 37% threshold was optimized to obtain the best possible correspondence between the aerial image and the binary map of the grass/forest boundary. Pixels are approximately 1 × 1 m, allowing for an individual tree-level resolution. This allowed the fine-tuning of the geolocation of the plot to obtain an almost perfect match between our individual tree map (trees within the plot) and the larger grass vs. forest map.

2.5. Data Analyses

As part of the first assessment on species composition and diversity conducted in the Ngel Nyaki Forest Dynamics Plot, we calculated abundances, basal area and Fisher's alpha diversity on a per hectare basis (i.e., within squares of 100 × 100 m, each a subset of the entire plot) for three diameter classes: ≥1 cm, ≥10 cm, and ≥30 cm. We used 13 table combinations of our plot for our calculations so that we had a total of 182 1-ha plots from which we calculated our per hectare diversity indices. We analyzed diversity at three taxonomic levels: species, genus, and family.

The four observed environmental preferences (% of surrounding grass, distance to grass, distance to stream, elevation) were compared with null expectations. The null expectations for each species were obtained after 999 constrained torus translations, an adaptation of the standard or unconstrained torus translation (see Appendix A for more details on standard (unconstrained) and constrained torus translations). During a standard torus translation, all individuals are translated together as a rigid cloud of points [39–41]. The spatial pattern of the species (the spatial aggregation) is respected, but the relationship with the habitat map (the grass vs. forest map, or the streams, or the elevation) is broken. All of our translations were constrained to only happen within the forest, never in the grass (Figure 3). We used the same constrained translations to assess significant associations with the grass, the streams and certain elevations. We developed constrained torus translations instead of unconstrained or standard torus translations across the entire plot because the latter predicted many individuals within the grass itself, just because a significant proportion of the Ngel Nyaki plot is covered in grass. This unrealistic expectation is a meaningless null hypothesis to test; if we did test it, it would result in all species being significantly far from the grass, even those with strong preferences for the forest edge.

After generating the null expectations, we evaluated the significance of the associations. If the observed x for the species was greater or lower than 95% of the expected (null) x, then the species was considered significantly associated to x. For example, if the observed median distance to grass of the individuals of a species was greater than 95% of the expected median distance after 999 torus translations, the species was deemed "significantly far from grass", i.e., a forest core specialist. In contrast, if it was lower than 95% of the expected median distances, the species was deemed "significantly close to grass". We followed the same approach with each of the other environmental variables considered. We used a Chi-Squared test to examine the relationship between species position on the gradient and seed size and dispersal mode. We grouped seeds into four size classes according to their weight: small (<0.6 g), medium (0.6–1.5 g), large (>1.5–3.0 g), and very large (>3.0 g). All the analyses were performed in R, version 3.3.3 [42] using customized functions or the CTFS R package [43].

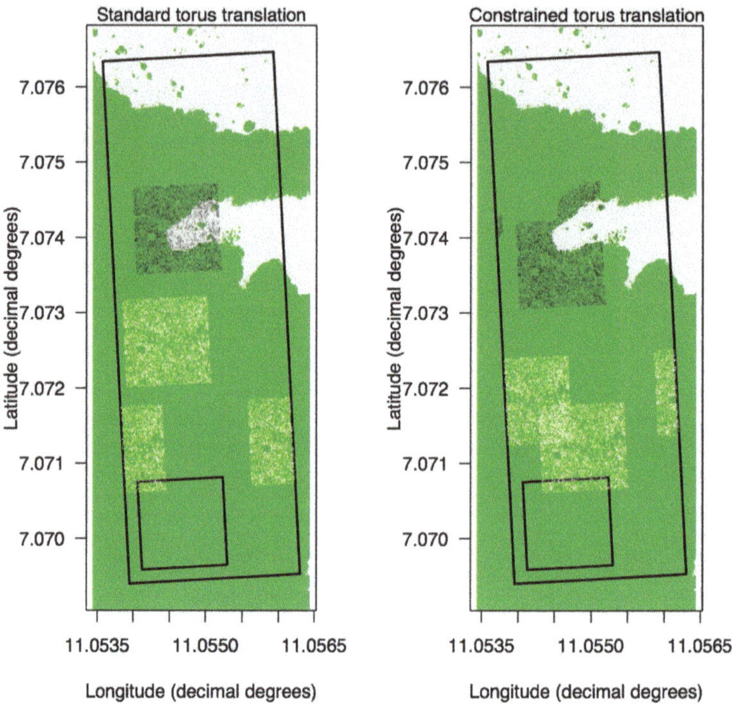

Figure 3. Comparison between the standard (unconstrained) torus translation (**A**) and the constrained version that we used in our study (**B**). The square polygon within the plot corresponds to the original position of a subset of stems, used as an example. Three random translations of this subset of stems are presented for comparison purposes. When the stems land within forest after the translation, both methods behave similarly (whitish points). When the stems land within grass, the constrained translation makes them "jump" again until they find forest, distorting to some degree the translated cloud of points (black points). The standard (unconstrained) translation does nothing in particular, since it is blind to the forest vs. grass distinction. This non-adaptive behavior of the standard torus translation contributes to the unrealistic null expectation of "many stems within the grass", that we wanted to avoid.

3. Results

3.1. Diversity and Abundance

A total of 41,031 trees were recorded with an average density of 2062 individuals/ha. The number of individuals ≥10 cm dbh and ≥30 cm dbh were 6931 (352/ha) and 1686 (85/ha), respectively (Table 1).

We recorded 105 species belonging to 47 families and 87 genera in the 20.28-ha plot. Rubiaceae, with nine species, was the most diverse family, followed by Euphobiaceae and Fabaceae with six species each, and Meliaceae with five species. Fisher's alpha at the 1-ha scale was 12.42 for all stems, 12.03 for stems ≥10 cm, and 9.55 for the largest diameter class (≥30 cm dbh) (Table 1). The most abundant species was *Garcinia smeathmannii*, an understory tree of the Clusiaceae family with 11,960 individuals (607.63/ha), representing approximately 30% of individuals ≥1 cm dbh (Table 2). This was followed by *Deinbollia pinnata* (Sapindaceae) with 3077 individuals (158.68/ha) and *Pleiocarpa pycnantha* (Apocynaceae) with 2142 individuals (107.51/ha).

Table 1. Tree abundance and diversity in the 20.28-ha Ngel Nyaki Plot, in three diameter at breast height (dbh) categories. Mean values and standard deviations are presented for per hectare estimates.

Variables	dbh ≥ 1 cm	dbh ≥ 10 cm	dbh ≥ 30 cm
Number of species	105	88 (83%)	61 (58%)
Number of genera	88	76 (86%)	56 (64%)
Number of families	47	43 (92%)	34 (72%)
Abundance	41,031	6931 (17%)	1686 (4%)
Total basal area (m^2)	553.49	498.24 (90.0%)	356.60 (64%)
Fisher's α per hectare	12.42 ± 1.51	12.03 ± 2.20	9.55 ± 2.74
Mean density per hectare (N/ha)	2061.86 ± 697.03	351.79 ± 112.74	85.04 ± 28.00
Mean basal area per hectare (m^2/ha)	27.52 ± 8.64	24.70 ± 7.87	17.62 ± 5.58

Table 2. The twenty most abundant woody plant species in the Ngel Nyaki plot, ranked by density. The rank corresponding to their basal area is indicated in parentheses.

Species	Family	Density (Trees/ha)	Basal Area (m^2/ha)
Garcinia smeathmanii (Planch. & Triana) Oliv.	Clusiaceae	607.63	1.20 (7)
Deinbollia pinnata Schumach. &Thonn.	Sapindaceae	158.68	0.49 (11)
Pleiocarpa pycnantha (K.Schum.) Stapf	Apocynaceae	107.51	0.48 (12)
Leptaulus zenkeri Engl.	Icacinaceae	69.19	1.25 (6)
Carapa oreophila Kenfack	Meliaceae	56.92	2.14 (3)
Chrysophyllum albidum G. Don	Sapotaceae	56.75	0.21 (18)
Sorindeia sp.	Anacardiaceae	53.10	0.25 (15)
Strombosia scheffleri Engl.	Olacaceae	51.13	2.76 (2)
Drypetes gossweilleri S. Moore	Putranjivaceae	50.42	0.63 (10)
Newtonia buchannani (Baker f.) G.C.C. Gilbert & Boutique	Fabaceae	47.95	0.87 (8)
Dicranolepis grandiflora Engl.	Thymelaeceae	47.54	0.04 (19)
Anthonotha noldeae (Rossberg) Exell and Hillc.	Fabaceae	46.24	3.42 (1)
Voacanga africana Stapf	Apocynaceae	44.52	0.22 (17)
Tabernaemontana contorta Stapf	Apocynaceae	40.38	0.23 (16)
Santiria trimera (Oliv.) Aubrév.	Burseraceae	40.36	1.76 (4)
Oxyanthus speciosus DC.	Rubiaceae	39.21	0.47 (13)
Psychotria peduncularis (Salisb.) Steyerm.	Rubiaceae	37.58	0.01 (20)
Macaranga occidentalis (Müll. Arg.) Müll. Arg.	Euphorbiaceae	34.90	0.84 (9)
Trichilia monadelpha (Thonn.) J.J. de Wilde	Meliaceae	28.18	1.40 (5)
Millettia conraui Harms	Fabaceae	23.04	0.27 (14)

3.2. Basal Area

Tree total basal area was 553.49 (27.52 ± 8.64 m^2/ha) of which 90% was contributed by large-diameter trees ≥30 cm dbh. The basal area was dominated by the family Fabaceae (96.67 m^2), followed by Meliaceae, Olacaceae, Burseraceae, Moraceae, Sapotaceae, and Icacinaceae. *Anthonotha noldeae* had the highest basal area, followed by *Strombosia schefleri, Carapa oreophila, Santiria trimera, Trichilia monadelpha,* and *Leptaulus zenkeri* (Table 2). The complete species list with total number of individuals and total basal area (m^2) for every species is found in Table S1.

3.3. Species Distribution and Preferences

The four gradients (distance from edge, distance to stream, elevation, % grass) evaluated in the plot were weakly correlated with each other (Figure 4). Individual trees at higher elevations were closer to the grass (Pearson's r = −0.46). The rest of the correlations were weaker than $|r|$ = 0.30 (Figure 4).

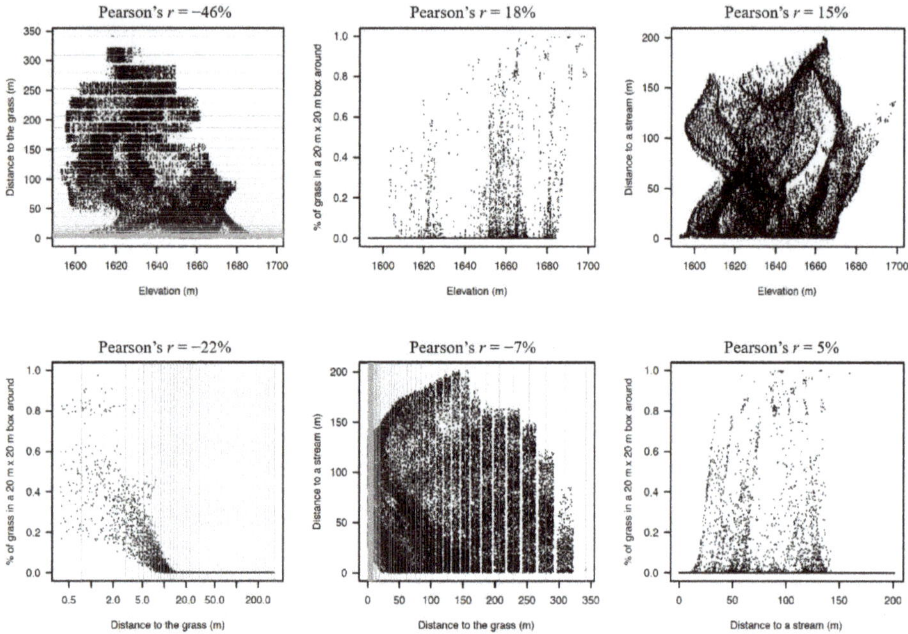

Figure 4. Distribution of individual trees along the four environmental gradients in the Ngel Nyaki plot. Individual trees occur across all gradients, therefore our analysis reflects the underlying correlation between individuals and measured local environmental variables. For computational reasons, the distance to the grass was measured with high resolution at short distances and coarser resolution at larger distances (the maximum peripheral error was constrained to <15%). The possible values for the distance to the grass are represented by thin grey lines in the figures; the points have been jittered around those possible values for visualization purposes.

The average individual values at the species level, however, showed stronger correlations, with some obvious elevational trends (Figure 5).

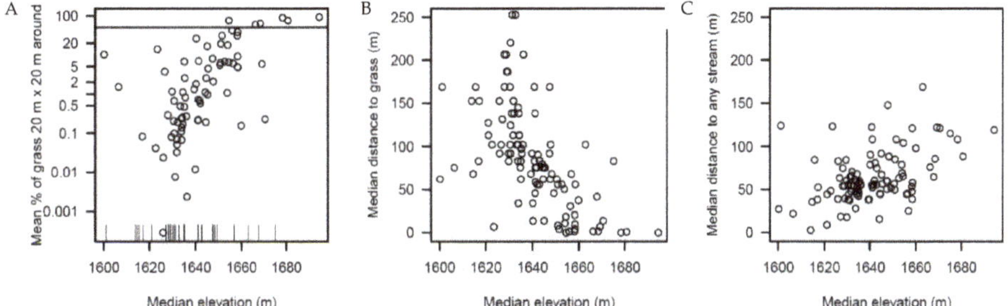

Figure 5. Environmental preferences of species in the Ngel Nyaki plot. Each point in the figure represents one species. (**A**) Median elevation of the individuals of each species vs. the mean % of grass pixels in a 20 × 20 m window around each individual of the species ($r = 0.57$). The vertical axis is in logarithmic scale; ticks in the horizontal axis represent species that have absolutely no grass around (zeroes cannot be represented in the logarithmic scale). The horizontal line at 50% separates four savannah species and two species from the rest. (**B**) Median elevation vs. median distance to the grass across all the individuals of each species ($r = -0.60$). (**C**) Median elevation vs. median distance to the closest stream across all the individuals of each species ($r = 0.45$).

Many of the average preferences for the species did not differ from the expected by the random translations, suggesting that many of the species in Ngel Nyaki do not respond to the measured environmental gradients (See Table S2). The more common species however, showed stronger preferences (Table 3). All of the 65 abundant species (those with at least 20 individuals in the plot) were surrounded by significantly more or less grass than the expected by the null model. Four of these were clearly savannah species (*Dombeya ledermannii*, *Entada abyssinica*, *Psorospermum aurantiacum*, *Rytigynia* sp.) and two species (*Clausena anisata* and *Guarea cedrata*) mostly occurred in areas with a high percentage of grass (>50%) even though they are not savannah species. More than half of the common species (35) were significantly close (23) or far (12) from the grass. Around 20% of the common species (14) were significantly associated with high (8) or low (6) elevations. However, in contrast, few common species (6) were significantly close (5) or far (1, Chionanthus) from the stream.

Our analysis investigating the association between species position on the gradient and seed size or dispersal mode showed no relationship between position on the gradient and seed size ($\chi = 4.36$, df $= 6$, $p = 0.63$) or dispersal mode ($\chi = 7.57$, df $= 6$, $p = 0.27$). For example, species significantly closer to the grassland matrix ("edge" species in Table 3) had a median seed size of 0.25 g, twice the median size of those species significantly far from the grassland matrix (0.13 g; "core" species in Table 3). Twenty five percent and 75% quantiles for species close and far from the grass are 0.13–0.34 and 0.04–0.27 g respectively.

Table 3. Associations between the distribution of the common species and four environmental gradients in the Ngel Nyaki plot (Nigeria). The numeric values correspond to the proportion of null values (medians across individuals, except in the case of % of surrounding grass, which are means) that are lower than the observed value. The columns "association" contain a short label for the ecological interpretation of the results: forest vs. savannah species, edge vs. core species (core = species that are significantly far from the grass), species that prefer low vs. high elevations, and species that prefer being close to the streams (wet) or far from them (dry). NA's mean that a given association does not deviate significantly from the expected by the null model. We include species guild, growth form, seed size, and dispersal mode (Growth-form categories of species were ET = emergent tree, CT = canopy tree, UT = understory tree, USh = understory shrub, ST = savannah tree, Sh = shrub (modified from [27]). Species shade tolerance guilds included P = pioneer, S = shade-tolerant, CP = Cryptic pioneer, NPLD= Non-pioneer light demanding, G = Grassland (modified from [35]). Dashes (-) signify insufficient data).

Species	% of Grass in 20 × 20 m Around		Distance to the Grass		Elevation		Distance to Streams		Guild	Growth Form	Seed Size (g)	Dispersal Mode
	Prop. of obs. > Null	Association	Prop. of obs. > Null	Association	Prop. of obs. > Null	Association	Prop. of obs. > Null	Association				
Leptaulus zenkeri	0	forest	1	core	0.159	NA	0.47	NA	S	CT	0.12	animal
Newtonia buchananii	0	forest	1	core	0.272	NA	0.157	NA	S	ET	0.14	wind
Chrysophylum albidum	0	forest	0.998	core	0.143	NA	0.818	NA	S	CT	0.35	animal
Voacanga africana	0	forest	0.992	core	0.154	NA	0.518	NA	S	UT	0.24	animal
Dicranolepis grandifolia	0	forest	0.992	core	0.157	NA	0.51	NA	S	UT	0.38	animal
Chionanthus africanus	0	forest	0.989	core	0.144	NA	0.988	dry	S	CT	-	animal
Drypetes gossweileri	0	forest	0.978	core	0.133	NA	0.028	wet	S	CT	0.01	animal
Oxyanthus speciosus	0	forest	0.975	core	0.326	NA	0.726	NA	S	UT	1.60	animal
Pavetta corymbosa	0	forest	0.974	core	0.479	NA	0.495	NA	S	USh	0.02	animal
Discoclaoxylon hexandrum	0	forest	0.963	core	0.119	NA	0.843	NA	S	USh	0.04	animal
Campylospermum flavum	0	forest	0.961	core	0.234	NA	0.035	wet	S	UT	-	animal
Dasylepis racemosa	0	forest	0.925	NA	0.680	NA	0.441	NA	S	UT	0.43	animal
Santiria trimera	0	forest	0.922	NA	0.024	low	0.085	NA	S	CT&UT	1.32	animal
Garcinia smeathmannii	0	forest	0.905	NA	0.141	NA	0.712	NA	S	UT	2.57	animal
Strombosia scheffleri	0	forest	0.872	NA	0.027	low	0.099	NA	S	CT	-	animal
Trichilia monadelpha	0	forest	0.862	NA	0.220	NA	0.046	wet	S	CT	0.50	animal
Sapium ellipticum	0	forest	0.841	NA	0.121	NA	0.014	wet	P	CT	-	animal
Diospyros mombuttensis	0	forest	0.743	NA	0.353	NA	0.913	NA	S	UT	0.68	animal
Kigellia africana	0	forest	0.606	NA	0.272	NA	0.397	NA	NPLD	CT	-	animal
Psychotria pedunculeris	0	forest	0.547	NA	0.263	NA	0.831	NA	P	USh	0.03	animal
Antidesma vogelianum	0	forest	0.515	NA	0.196	NA	0.675	NA	P	USh	0.03	animal
Trilepisium madagascariense	0	forest	0.451	NA	0.479	NA	0.640	NA	-	CT	0.97	animal
Symphonia globulifera	0	forest	0.444	NA	0.056	NA	0.409	NA	S	CT	0.45	animal
Polyscias fulva	0	forest	0.442	NA	0.228	NA	0.536	NA	P/CP	UT	0.01	animal
Celtis gomphophylla	0	forest	0.430	NA	0.154	NA	0.662	NA	P/CP	UT	0.01	animal
Carapa oreophila	0	forest	0.411	NA	0.032	low	0.063	NA	NPLD	UT	18.51	gravity
Xymalos monospora	0	forest	0.398	NA	0.667	NA	0.172	NA	S	UT	0.17	animal
Parkia filicoidea	0	forest	0.355	NA	0.048	low	0.939	NA	NPLD	CT	0.77	animal
Mallotus oppositifolius	0	forest	0.317	NA	0.418	NA	0.115	NA	S	USh	0.02	animal
Zanthoxylum leprieurii	0	forest	0.215	NA	0.803	NA	0.540	NA	CP	UT	0.02	animal
Ficus sur	0	forest	0.211	NA	0.351	NA	0.165	NA	Strangler	Strangler	-	animal

Table 3. *Cont.*

Species	% of Grass in 20 × 20 m Around		Distance to the Grass		Elevation		Distance to Streams		Guild	Growth Form	Seed Size (g)	Dispersal Mode
	Prop. of obs. > Null	Association	Prop. of obs. > Null	Association	Prop. of obs. > Null	Association	Prop. of obs. > Null	Association				
Leea guineensis	0	forest	0.202	NA	0.620	NA	0.569	NA	CP	USh	0.04	animal
Psychotria succulenta	0	forest	0.189	NA	0.358	NA	0.855	NA	P	USh	-	animal
Trema orientalis	0	forest	0.169	NA	0.011	low	0.236	NA	P/CP	UT	0.01	animal
Nuxia congesta	0	forest	0.167	NA	0.017	low	0.382	NA	P	UT	<0.001	animal
Albizia gummifera	0	forest	0.136	NA	0.980	high	0.664	NA	P/CP	CT	0.55	wind
Entandophragma angolense	0	forest	0.133	NA	0.063	NA	0.885	NA	NPLD	ET	1.73	wind
Macaranga occidentalis	0	forest	0.107	NA	0.327	NA	0.008	wet	P	UT	0.01	animal
Deinbollia pinnata	0	forest	0.104	NA	0.855	NA	0.841	NA	S	UT	0.62	animal
Pouteria altissima	0	forest	0.103	NA	0.833	NA	0.226	NA	NPLD	ET	3.19	animal
Ficus lutea	0	forest	0.051	NA	0.729	NA	0.213	NA	Strangler	Strangler	-	animal
Pleiocarpa pycnantha	0	forest	0.030	edge	0.892	NA	0.530	NA	S	UT	0.25	animal
Beilschmiedia mannii	0	forest	0.013	edge	0.816	NA	0.464	NA	S	UT	1.16	animal
Isolona sp.	0	forest	0.010	edge	0.870	NA	0.649	NA	S	UT	1.62	animal
Millettia conraui	0	forest	0.007	edge	0.796	NA	0.326	NA	P/CP	UT	0.20	ballistic
Warneckea cinnamomoides	0	forest	0.005	edge	0.664	NA	0.623	NA	P/CP	USh	-	animal
Rauvolfia vomitoria	0	forest	0.002	edge	0.520	NA	0.829	NA	P/CP	USh	0.07	animal
Rothmania urcelliformis	0	forest	0.002	edge	0.999	high	0.551	NA	S	UT	0.31	animal
Eugenia gilgii	0	forest	0.001	edge	0.806	NA	0.457	NA	P	UT	0.11	animal
Ritchiea albersii	0	forest	0.001	edge	0.996	high	0.774	NA	S	UT	-	animal
Clausena anisata	0	forest	0	edge	0.966	high	0.783	NA	P/CP	USh	0.07	animal
Sorindeia sp.	0	forest	0	edge	0.754	NA	0.713	NA	NPLD	CT	0.30	animal
Tabanaemontana contorta	0	forest	0	edge	0.971	high	0.922	NA	P/CP	UT	0.35	animal
Anthonotha noldeae	0	forest	0	edge	0.968	high	0.696	NA	P/CP	CT	6.57	ballistic
Bridelia speciosa	0	forest	0	edge	0.510	NA	0.847	NA	P/CP	UT	-	wind
Ficus sp.	0	forest	0	edge	0.945	NA	0.440	NA	Strangler	Strangler	-	animal
Psychotria umbellata	0	forest	0	edge	0.602	NA	0.923	NA	P	USh	-	animal
Guarea cedrata	0	forest	0	edge	0.300	NA	0.896	NA	P/CP	UT	-	animal
Psorospermum aurantiacum	1	savanna	0	edge	0.751	NA	0.907	NA	G	Sh	0.01	animal
Entada abyssinica	1	savanna	0	edge	0.999	high	0.852	NA	G	ST	0.25	wind
Dombeya ledermannii	1	savanna	0	edge	0.840	NA	0.467	NA	G	ST	-	ballistic
Rytigynia sp.	1	savanna	0	edge	0.797	NA	0.804	NA	G	Sh	-	animal
Rubiaceae unidentified	0	forest	0.998	core	0.359	NA	0.518	NA	S	UT	-	animal
Unidentified	0	forest	0.033	edge	0.475	NA	0.553	NA	-	-	-	
Unidentified	0	forest	0.002	edge	0.997	high	0.350	NA	P	U	-	

4. Discussion

Our study is the first detailed inventory of a West African montane forest that explores species distributions in relation to local environmental gradients. Our data are from the first census of the Ngel Nyaki Smithsonian Forest Global Earth Observatory plot.

4.1. Diversity and Composition of the Woody Plant Species

Within our total plant count of 41,031 trees with dbh ≥1 cm we recorded 105 species from 87 genera and 47 families. This is relatively low diversity compared with the results from previous studies in other tropical Afromontane forests. For example, Hamilton and Bensted-Smith [44], who recorded only trees with dbh >10 cm counted 104 species from the sub-montane forest of the East Usambara Mountains in East Africa. Dowsett-Lemaire, et al. [45], similarly including only larger sized trees recorded 232 species over 140 km² in northern Malawi, with less diversity south of 14° S; e.g., 94 species in a forest of 11.3 km² in Central Malawi. This drop in species diversity in Southern Malawi was due to range limits and several species 'skipping' southern Malawi and reappearing farther south [45]. However, in concordance with some Afromontane forests, e.g., forests in the Bonga region in East Africa [46], a high proportion (90%) of the genera at Ngel Nyaki were each represented by a single species and the most common families were Rubiaceae and Fabaceae [47,48].

Of the total woody plants encountered, 83% were small stems <10 cm dbh, as is typical in African [47–49] and non-African old-growth forests [50]. The African lowland forests of Korup (Cameroon), Ituri (Democratic Republic of the Congo), and Rabi (Gabon) contain even higher proportions of small diameter trees (92.3%, 94.8%, 93.6%, respectively). The same is observed in other tropical forests: BCI (Panama) (90.1%), Sinharaja (Sri Lanka) (91.8%), Lambir and Pasoh (both in Malaysia) (90.8% in each), and Yasuni, (Ecuador) (88.5%) [48,50].

The number of individuals of species in the plot varied considerably, from 11 singletons (e.g., *Terminalia ivorensis* and *Olax* sp.) to the extremely abundant *Garcinia smeathmannii* (Clusiaceae) with about 12,000 individuals (607 trees/ha). *Garcinia smeathmannii*, which makes up about one-third of the total tree abundance in the plot, is also the third most abundant species in the Rabi plot (Gabon), with 304 trees/ha [48]. *Anthonotha noldeae* (Fabaceae), an Afromontane endemic and one of the most common canopy species in Ngel Nyaki forest, showed the highest dominance if measured by basal area. This is of interest because locally, *A. noldeae* is only found in Ngel Nyaki forest and Cabbal Shirgu, and is absent from the rest of the Gotel mountains in Nigeria [51] and Mt Oku in neighbouring Cameroon [52]. We recorded 607 trees/ha of our most abundant species *G. smeathmannii* which is comparable to the abundance of 684 trees/ha of *Dichostemma glaucescens* in Rabi (Gabon) [48] and 535 trees/ha of *Phyllobotryon spathulatum* in Korup (Cameroon) [47]. About 39% of the species in the plot had <1 tree/ha, which when compared to other ForestGEO plots is similar to the 45% in Korup (Cameroon) and BCI (Panama), 34% in Pasoh (Malysia) and 40% in Yasuni (Equador), but lower than in the monodominant Ituri forest (Democratic Republic of Congo) where 72% of the species there are represented by <1 individual per hectare [47,50]. Among large diameter trees (≥10 cm dbh), *Strombosia schefleri* (Erythropalaceae) was the most abundant species with 35.3 trees/ha which is very similar to the abundance of *Trichilia tuberculata* (Meliaceae), the commonest large tree in BCI (Panama) with a reported abundance of 35 trees/ha. Although Ngel Nyaki is less species diverse when compared with many tropical forests of Africa, it comprised species both widespread across Africa and Afromontane endemics or near-endemics *sensu* White [2].

4.2. The Role of Environmental Variables

The main aim of this study was to look for evidence of local environmental factors influencing species distributions. In terms of core to edge gradients, we tested the hypothesis that species sorting from the core forest to the forest edge should result from the progressive encroachment of forest into the grassland matrix via dispersal of small seeded, light tolerant species [26]. We found partial support for this hypothesis in that 40% of all tree species and 50% of the abundant species did show a clear preference for either the edge/grassland habitat or the forest core. However, we found no association between position on the gradient and seed size or dispersal mode, which suggests that factors other than dispersal are influencing species distributions (see below).

Many of the species strongly associated with edge were expected; e.g., *Entada abyssinica, Psorospermum aurantiacum, Rytiginia* sp., and *Dombeya ledermannii* are all species restricted to grassland or the very edge of the forest, where light is always abundant. Other edge associated species such as *Bridelia speciosa, Rauvolvia vomitoria, Millettia conraui*, and *Eugenia gilgii* are also fire resistant, small to medium-sized pioneers, but occur in the understory as well as on the forest edge. Jackson [53] noted a similar guild in the Imatong Mountains of Sudan, and Adie, Kotze, and Lawes [26] in the Drakensberg montane forest of South Africa. That these light demanding species occur in the forest understory indicates their 'cryptic pioneer' [35,54] status. That is, they need light to germinate and establish but as they increase in size, they become more shade tolerant and can survive in the understory as adults for many years. Hawthorne [35] suggested that in Ugandan forests these shade tolerant traits may be adaptive and not necessarily present throughout a species range, which would be worth investigating. Of note is that in our study site the very common canopy legume tree, *Anthonotha noldeae* also fits the description of a cryptic pioneer. *Anthonotha*'s abundance on the forest edge and its likely nitrogen-fixing properties [55] suggests it is an important component of the forest flora, potentially playing a key role in forest expansion.

Species strongly associated with the core forest were all shade tolerant, forest species and included the emergent wind dispersed, *Newtonia buchannii*, and animal dispersed canopy species such as *Drypetes gossweileri* and *Chrysophyllum albidum*. Understory shade species included *Voacanga africana* and *Dicranolepis grandifolia*. Most of these core forest species are widespread across Africa and in Nigeria are a components of forests up to at least 1800 m in elevation [27].

Against our hypothesis for a core to edge gradient we found that almost 50% of species with ≥20 individuals in the plot (seven of which were in the 20 most abundant species in the plot), occurred throughout the forest, with no statistical preference for edge or core. Together, these species included almost the entire range of species guilds. For example, the understory tree species *Carapa oreophila, Celtis gomphophylla*, and *Mallotus oppositifolius* are 'cryptic pioneers', present deep within the forest as adults but showing no evidence of regeneration [56]. In contrast, species such as *Parkia filicoidea, Entandrophragma angolense* and *Pouteria altissima* are 'Non-Pioneer Light Demanding' [35] in that they need shade to germinate and establish, but as adults are light demanding. Shade tolerant understory species with no associations with either core or edge include *Dasylepis racemosa, Diospyros monbuttensis*, and *Santiria trimera*.

In contrast to our expectations, we found that species closer to the edge had, on average, larger seed sizes than the core species. So that unlike the situation in the South African temperate montane forests where small seeds are obviously key to forest success, we found that in West Africa, larger-seeded and fire-adapted species seem to dominate the edges of each forest patch. Other adaptations, like nitrogen-fixing nodules in *Milletia* [57] and likely *A. noldeae* [55] may help in seedling establishment and subsequent survival in the grassland matrix. In general, how these forest patches grow and function as species refugia during times of forest expansion [26] are exciting questions and opportunities for further research.

While the grassland to forest gradient was the steepest in our analysis, we also explored other potentially relevant gradients to further explain the distribution of tree species at Ngel Nyaki. For example, elevation gradients often play a key role in driving functional diversity [58]. We found 14 species were significantly associated with either low or high elevation within the plot. Among those associated with high elevation, *Anthonotha noldeae, Clausena anisata, Entada abysinnica, Ritchea albersii, Rothmannia urcelliformis*, and *Tabanaemontana contorta* were also associated with grass/forest edge and distance to grass, which makes sense considering that the grasslands are at high elevation in our plot.

Another gradient explored was distance to stream (a proxy for soil moisture) but this was only weakly associated to species distribution in Ngel Nyaki, with only 6 of our 65 abundant species showing significant associations to distance from stream ("wet" or "dry" species in Table 3). This was unexpected given that water availability is well known to play a key role in governing local patterns of plant distributions in many ecosystems [59]. For example, in the Barro Colorado Island plot, Comita and Engelbrecht [60] reported significant associations between species distributions and water availability. In contrast however, and more like us, Scalley, et al. [61] in the Luquillo wet montane forest plot in Puerto Rico found no association between the spatial distribution of tree species and stream locations, but rather, that stem density and species richness was highest in areas closest to streams.

Our findings may, to some extent, corroborate those of Adie, Kotze, and Lawes [26] who suggest that in the South African Drakensberg mountains, fire, coupled with harsh abiotic conditions have filtered Afrotemperate forest tree composition resulting in a prevalence of species traits such as small, bird-dispersed fruits and fire resistant bark across the forest. Such filtering could perhaps explain the presence of the long lived 'cryptic' pioneers *Polyscias fulva* and *Trema orientalis* throughout Ngel Nyaki forest. However, in Nigerian montane forests there is a range of species guilds and seed sizes; while some may reflect species filtering through the matrix, others are shade tolerant, late succession old forest species such as *Santiria trimera* [61]. Regarding *S. trimera*, both the stilt root and non-stilt root morphs occur at our study site. The separation between the morphs is deep rooted in the *Santiria* phylogenetic tree [62], therefore their presence at Ngel Nyaki perhaps suggests that Ngel Nyaki forest is ancient and was a forest refugia during periods of Pleistocene drying. Population genetic analysis following that of Koffi, Hardy, Doumenge, Cruaud and Heuertz [62] on Ngel Nyaki *S. trimera* and other forest tree species would help shed light on this. However, that Ngel Nyaki forest is the most species rich forest on the Mambilla Plateau [45] with several rare species [25] and some even new to science [63], and that *A. noldeae* occurs here in such abundance but almost nowhere else in West Africa, may again support the idea that Ngel Nyaki was a Pleistocene refuge. Moreover Brailsford [64] recorded high levels of chloroplast haplotype diversity at Ngel Nyaki as compared with other forest fragments on the Mambilla Plateau. Of course alternative explanations exist for the high diversity and again, molecular analyses would help shed light on this.

5. Conclusions

Our study provides the first detailed insight on patterns of species and trait guild composition in a West African montane forest. The Ngel Nyaki forest is similar to many other tropical and Afrotropical forests in terms of species diversity and dominance patterns. The gradient from forest edge to core is a key dimension of the niche as perceived by plants. Overall, ~40% of the individuals belong to species with significant preferences towards edges or forest core areas, which in turn represent ~45% of the local woody diversity. While dispersal abilities play a role in forest expansion, our data did not show any association between species position on the gradient and dispersal mode. Further research is needed to disentangle the effect of changes in forest patch size and environmental preferences of the species to make more accurate predictions of how Afromontane forests will behave under changing fire regimes and human disturbance in the region.

Supplementary Materials: The following are available online at http://www.mdpi.com/1424-2818/12/1/30/s1, Table S1, Total number of individuals and total basal area (m^2) for each of the 105 species of trees with dbh ≥1cm recorded in the Ngel Nyaki 20.28-ha plot. Table S2, Associations between the distribution of the all species and four environmental gradients in Ngel Nyaki plot (Nigeria). Full census data are available upon reasonable request from the ForestGEO data portal, http://ctfs.si.edu/datarequest/.

Author Contributions: Conceptualization, I.A., G.A., D.K., and H.C.; data curation, I.A., D.K., and H.C.; formal analysis, I.A. and G.A.; funding acquisition, H.C.; investigation, I.A., G.A., D.K., and H.C.; methodology, I.A., G.A., D.K., and H.C.; project administration, H.C.; software, G.A.; supervision, D.K. and H.C.; visualization, G.A.; writing—original draft, I.A.; writing—review and editing, I.A., G.A., D.K., and H.C. All authors have read and agreed to the published version of the manuscript.

Funding: The setting up and enumeration of the ForestGEO plot in Ngel Nyaki, Nigeria was majorly funded by a donation from Retired General T.Y. Danjuma to the Nigerian Montane Forest Project and other smaller donations from Forest Global Earth Observatory (ForestGEO) of the Smithsonian Tropical Research Institute, Chester Zoo, England and the A.G. Leventis Foundation.

Acknowledgments: This study was made possible by the generous donation made by Retired General T.Y. Danjuma to the Nigerian Montane Forest Project (NMFP) through H. Chapman. We also thank Forest Global Earth Observatory (ForestGEO) of the Smithsonian Tropical Research Institute for the support towards the plot census. We also wish to thank Chester Zoo, England, and A.G. Leventis Foundation for their financial assistance to the NMFP. I. Abiem also received a PhD scholarship from the University of Jos, Nigeria. We wish to thank the staff of NMFP especially the ForestGEO team who carried out the tree census. We also wish to thank S. Russo for her enlightening suggestions on the scope and direction of this research and for her extremely generous mentorship to IA during the ForestGEO workshops. The idea for this manuscript was put together at the 2017 ForestGEO workshop supported by the US National Science Foundation (NSF) grant to S. J. Davies. GA was supported while conducting this research as part of the Next Generation Ecosystem Experiments-Tropics, funded by the U.S. Department of Energy, Office of Science, Office of Biological and Environmental Research.

Conflicts of Interest: The authors declare no conflict of interest.

Appendix A

Appendix A.1 Torus Translations

The goal of a torus translation is to break the relationship between a cloud of points (e.g., the individuals of a given species) and the space and consequently, any biotic/abiotic gradient that happens in the space.

Torus translation tests are routinely used in analysis of plant-habitat associations [40–42]. The rationale underlying torus translation tests is to contrast the observations with the expectations under the assumption of lack of relationship between a given species and the habitat. The expectation is generated by breaking the spatial relationship between the habitat map and the tree map.

While a simple approach to break the relationship between, e.g., trees and the space they occupy would be to randomize the location of each tree independently, this would treat each tree as an independent observation, which is incorrect. To avoid this problem, a torus translation is a translation of the entire cloud of points (the entire map of trees). In particular, it is a translation in a torus: when making a random "jump", things that "disappear" by the right appear by the left, things that "disappear" by the north appear by the south, and so on.

Appendix A.2 Translating the Map vs. the Individuals

Previous implementations of the method, at least in large Forest GEO plots similar to Ngel-Nyaki, have translated a categorical habitat map in the torus. By doing so: (1) the relationship between individual trees and habitat is broken, and (2) the spatial aggregation of the individual trees is respected. Consequently, one can test whether a given species has more/less individuals within a given habitat than expected.

Exactly the same can be done by translating the trees themselves (i.e., translating the cloud of points that represent the individual trees. The result is the same: a *relative* translation between the map and the trees. Here, we decided to translate the points, not the map. We considered it more intuitive to think of variables like "distance to the stream", "% of grass surrounding", etc., as individual-level properties, more than variables in a map.

Appendix A.3 Constrained Translations

The point of making the torus translations is to generate a null expectation. Something with which to contrast the observations. The translations define a null hypothesis, and the null hypothesis must make sense. In our case, we were particularly interested in the forest vs. grass distinction. We wanted to know whether some species were closer to the grass than others (i.e., closer to the edge of the forest). The Ngel Nyaki plot itself, however, is not completely covered by forest. It has a substantial amount of grass within it. If we translated points to forest or grass indistinctly, we would end with many trees within the grass. This would generate the (erroneous) expectation of all the species being strongly associated with grass. By comparing that expectation with the observations, we would conclude that all the species are significantly far from the grass. Even edge specialists.

To avoid that we run constrained translations in the torus. Like standard torus translations, a constrained translation makes the trees jump together (as a rigid cloud of points) randomly in the torus. Unlike the standard torus translations, we added the constraint of never arriving in grass. Jumps were forced to be sufficiently long that they always arrived in forest rather than grass, while at the same time respecting as much as possible the relative position of the translated points. One way of thinking about this is as a non-perfect torus, narrower in places with more grass, and wider in places with more forest. The torus is still a torus, topologically speaking. Another way to think about this is to imagine a map with invaginations in the forbidden areas, on top of which a regular grid of coordinates is placed (with no invaginations), to serve as a reference for the translations.

In summary, a constrained torus translation is a translation of a cloud of points that respects, as much as possible, the relative position of the individuals. Species that are strongly aggregated have expectations composed of tight clouds of points, and vice versa. We inspected visually many translations and we concluded that the result was satisfactory for our purposes.

References

1. Linder, H.P. The evolution of African plant diversity. *Front. Ecol. Evol.* **2014**, *2*, 38. [CrossRef]
2. White, F. *The Vegetation of Africa, a Descriptive Memoir to Accompany the UNESCO/AETFAT/UNSO Vegetation Map of Africa*; UNESCO: Paris, France, 1983; 356p.
3. White, F. The afromontane region. In *Biogeography and Ecology of Southern Africa*; Springer: Berlin/Heidelberg, Germany, 1978; pp. 463–513.
4. Gehrke, B.; Linder, H.P. Species richness, endemism and species composition in the tropical Afroalpine flora. *Alp. Bot.* **2014**, *124*, 165–177. [CrossRef]
5. Linder, H.P.; Lovett, J.; Mutke, J.M.; Barthlott, W.; Jürgens, N.; Rebelo, T.; Küper, W. A numerical re-evaluation of the sub-Saharan phytochoria of mainland Africa. *Biol. Skr.* **2005**, *55*, 229–252.
6. Burgess, N.D.; Balmford, A.; Cordeiro, N.J.; Fjeldsa, J.; Küper, W.; Rahbek, C.; Sanderson, E.W.; Scharlemann, J.P.; Sommer, J.H.; Williams, P.H. Correlations among species distributions, human density and human infrastructure across the high biodiversity tropical mountains of Africa. *J. Biol. Conserv.* **2007**, *134*, 164–177. [CrossRef]
7. Cordeiro, N.J.; Burgess, N.D.; Dovie, D.B.; Kaplin, B.A.; Plumptre, A.J.; Marrs, R. Conservation in areas of high population density in sub-Saharan Africa. *Biol. Conserv.* **2007**, *134*, 155–163. [CrossRef]
8. Meadows, M.; Linder, H. Special Paper: A Palaeoecological perspective on the origin of afromontane grasslands. *J. Biogeogr.* **1993**, *20*, 345–355. [CrossRef]
9. Spracklen, D.; Righelato, R. Tropical montane forests are a larger than expected global carbon store. *Biogeosciences* **2014**, *11*, 2741–2754. [CrossRef]
10. Schröter, D.; Cramer, W.; Leemans, R.; Prentice, I.C.; Araújo, M.B.; Arnell, N.W.; Bondeau, A.; Bugmann, H.; Carter, T.R.; Gracia, C.A. Ecosystem service supply and vulnerability to global change in Europe. *Science* **2005**, *310*, 1333–1337. [CrossRef]
11. Nadkarni, N.M.; Wheelwright, N.T. *Monteverde: Ecology and Conservation of a Tropical Cloud Forest*; Oxford University Press: Oxford, UK, 2000.
12. Chapman, H.M.; Olson, S.M.; Trumm, D. An assessment of changes in the montane forests of Taraba State, Nigeria, over the past 30 years. *Oryx* **2004**, *38*, 282–290. [CrossRef]
13. Lézine, A.-M.; Izumi, K.; Kageyama, M.; Achoundong, G. A 90,000-year record of Afromontane forest responses to climate change. *Science* **2019**, *363*, 177–181. [CrossRef]
14. Lawes, M.J. The Distribution of the Samango Monkey (Cercopithecus mitis erythrarchus Peters, 1852 and Cercopithecus mitis labiatus I. Geoffroy, 1843) and Forest History in Southern Africa. *J. Biogeogr.* **1990**, *17*, 669–680. [CrossRef]

15. White, G.C.; Burnham, K.P.; Anderson, D.R. Advanced features of Program Mark. In *Wildlife, Land, and People: Priorities for the 21st Century*; Field, R., Warren, R.J., Okarma, H., Sievert, P.R., Eds.; The Wildlife Society: Bethesda, MD, USA, 2001.

16. Lézine, A.-M.; Assi-Kaudjhis, C.; Roche, E.; Vincens, A.; Achoundong, G. Towards an understanding of West African montane forest response to climate change. *J. Biogeogr.* **2013**, *40*, 183–196. [CrossRef]

17. Eeley, H.A.; Lawes, M.J.; Piper, S. The influence of climate change on the distribution of indigenous forest in KwaZulu-Natal, South Africa. *J. Biogeogr.* **1999**, *26*, 595–617. [CrossRef]

18. Dowsett-Lemaire, F.; Dowsett, R.J.; Dyer, M. *Important Bird Areas in Africa and Associated Islands*; Pisces Publications and BirdLife International: Cambridge, UK, 2001.

19. Lebamba. Forest-savannah dynamics on the Adamawa plateau (Central Cameroon) during the "African humid period" termination: A new high-resolution pollen record from Lake Tizong. *Rev. Palaeobot. Palynol.* **2016**, *235*. [CrossRef]

20. Venter, Z.S.; Cramer, M.D.; Hawkins, H.J. Drivers of woody plant encroachment over Africa. *Nat. Commun.* **2018**, *9*, 2272. [CrossRef]

21. Müller, S.C.; Overbeck, G.E.; Pfadenhauer, J.; Pillar, V.D. Woody species patterns at forest–grassland boundaries in southern Brazil. *Flora Morphol. Distrib. Funct. Ecol. Plants* **2012**, *207*, 586–598. [CrossRef]

22. Gignoux, J.; Konaté, S.; Lahoreau, G.; Le Roux, X.; Simioni, G. Allocation strategies of savanna and forest tree seedlings in response to fire and shading: Outcomes of a field experiment. *Sci. Rep.* **2016**, *6*, 38838. [CrossRef]

23. Kotze, D.J.; Lawes, M.J. Viability of ecological processes in small Afromontane forest patches in South Africa. *J. Austral Ecol.* **2007**, *32*, 294–304. [CrossRef]

24. Andela, N.; Morton, D.; Giglio, L.; Chen, Y.; Van Der Werf, G.; Kasibhatla, P.; DeFries, R.; Collatz, G.; Hantson, S.; Kloster, S. A human-driven decline in global burned area. *Science* **2017**, *356*, 1356–1362. [CrossRef]

25. Van Langevelde, F.; Van De Vijver, C.A.; Kumar, L.; Van De Koppel, J.; De Ridder, N.; Van Andel, J.; Skidmore, A.K.; Hearne, J.W.; Stroosnijder, L.; Bond, W.J. Effects of fire and herbivory on the stability of savanna ecosystems. *Ecology* **2003**, *84*, 337–350. [CrossRef]

26. Adie, H.; Kotze, D.J.; Lawes, M.J. Small fire refugia in the grassy matrix and the persistence of Afrotemperate forest in the Drakensberg Mountains. *Sci. Rep.* **2017**, *7*, 6549. [CrossRef] [PubMed]

27. Chapman, J.; Chapman, H. *The Forests of Taraba and Adamawa States, Nigeria an Ecological Account and Plant Species Checklist*; University of Canterbury: Christchurch, New Zealand, 2001; p. 221.

28. Beck, J.; Chapman, H. A population estimate of the endangered chimpanzee Pan troglodytes vellerosus in a Nigerian montane forest: Implications for conservation. *Oryx* **2008**, *42*, 448. [CrossRef]

29. Nigerian Montane Forest Project. Available online: https://www.canterbury.ac.nz/afromontane/ (accessed on 6 November 2019).

30. Borokini, T. A Systematic Compilation of IUCN Red-listed Threatened Plant Species in Nigeria. *Int. J. Environ. Sci.* **2014**, *3*, 104–133.

31. Thia, J.A. The Plight of Trees in Disturbed Forest: Conservation of Montane Trees, Nigeria. Master's Thesis, University of Canterbury, Christchurch, New Zealand, 2014.

32. Condit, R. *Tropical Forest Census Plots: Methods and Results from Barro Colorado Island, Panama and A Comparison with Other Plots*; Springer Science & Business Media: Berlin, Germany, 1998.

33. Barcode of Life Data System. Available online: https://www.boldsystems.org/ (accessed on 6 November 2019).

34. The World's Most Detailed Globe. Available online: https://www.google.com/earth/ (accessed on 6 November 2019).

35. Hawthorne, W. Ecological profiles of Ghanaian forest trees. In *Tropical Forestry Papers*; Oxford Forestry Institute: Oxford, UK, 1995; Volume 29, pp. 1–345.

36. Hawthorne, W.D. Holes and the sums of parts in Ghanaian forest: Regeneration, scale and sustainable use. *Proc. R. Soc. Edinb. Sect. B Biol. Sci.* **1996**, *104*, 75–176. [CrossRef]

37. Global Plants. Available online: https://plants.jstor.org/ (accessed on 6 November 2019).

38. Barnes, A.D.; Chapman, H.M. Dispersal traits determine passive restoration trajectory of a Nigerian montane forest. *Acta Oecol.* **2014**, *56*, 32–40. [CrossRef]

39. Harms, K.E.; Condit, R.; Hubbell, S.P.; Foster, R.B. Habitat associations of trees and shrubs in a 50-ha Neotropical forest plot. *J. Ecol.* **2001**, *89*, 947–959. [CrossRef]

40. Wiegand, T.; Moloney, K.A. *Handbook of Spatial Point-Pattern Analysis in Ecology*; CRC: Boca Raton, FL, USA, 2013.

41. Arellano, G.; Medina, N.G.; Tan, S.; Mohamad, M.; Davies, S.J. Crown damage and the mortality of tropical trees. *New Phytol.* **2019**, *221*, 169–179. [CrossRef]

42. RCoreTeam. *R: A Language and Environment for Statistical Computing*; Version 3.3.3; R Foundation for Statistical Computing: Vienna, Austria, 2017.

43. The CTFS R Package. Available online: http://ctfs.si.edu/Public/CTFSRPackage/ (accessed on 6 November 2019).

44. Hamilton, A.C.; Bensted-Smith, R. *Forest Conservation in the East Usambara Mountains, Tanzania*; IUCN: Dar es Salaam, Tanzania, 1989; Volume 15.

45. Dowsett-Lemaire, F. The flora and phytogeography of the evergreen forests of Malawi I: Afromontane and mid-altitude forests. *Bull. Jard. Bot. Natl. Belg./Bull. Natl. Plantentuin Belg.* **1989**, *59*, 3–131. [CrossRef]

46. Schmitt, C.B.; Denich, M.; Demissew, S.; Friis, I.; Boehmer, H.J. Floristic diversity in fragmented Afromontane rainforests: Altitudinal variation and conservation importance. *Appl. Veg. Sci.* **2010**, *13*, 291–304. [CrossRef]

47. Kenfack, D.; Thomas, D.W.; Chuyong, G.B.; Condit, R. Rarity and abundance in a diverse African forest. *Biodivers. Conserv.* **2007**, *16*, 2045–2074. [CrossRef]

48. Memiaghe, H.; Lutz, J.; Korte, L.; Alonso, A.; Kenfack, D. Ecological importance of small-diameter trees to the structure, diversirty and biomass of a Tropical Evergreen Forest at Rabi, Gabon. *PLoS ONE* **2016**, *11*, e0154988. [CrossRef] [PubMed]

49. Makana, J.; Ewango, C.; McMahon, S.; Thomas, S.; Hart, T.; Condit, R. Demography and biomass change in monodominant and mixed old-growth forest of the Congo. *J. Trop. Ecol.* **2011**, *27*, 447–461. [CrossRef]

50. Valencia, R.; Foster, R.B.; Villa, G.; Condit, R.; Svenning, J.C.; Hernandez, C.; Romoleroux, K.; Losos, E.; Magard, E.; Balslev, H. Tree species distributions and local habitat variation in the Amazon: Large forest plot in eastern Ecuador. *J. Ecol.* **2004**, *92*, 214–229. [CrossRef]

51. Condit, R.; Ashton, P.; Baslev, H.; Brokaw, N.; Bunyavejchewin, S.; Chuyong, G.; Co, L.; Dattaraja, H.; Davies, S.; Esufali, S.; et al. Tropical tree alpha-diversity: Results from a worldwide network of large plots. *Biol. Skr.* **2005**, *55*, 565–582.

52. Beavon, M.A.; Chapman, H.M. Andromonoecy and high fruit abortion in Anthonotha noldeae in a West African montane forest. *Plant Syst. Evol.* **2011**, *296*, 217–224. [CrossRef]

53. Jackson, J.K. The vegetation of the Imatong Mountains, Sudan. *J. Ecol.* **1956**, 341–374. [CrossRef]

54. Grubb, P.J. Rainforest dynamics: The need for new paradigms. In *Tropical Rainforest Research—Current Issues*; Edwards, D.S., Booth, W.E., Choy, S.C., Eds.; Springer: Berlin/Heidelberg, Germany, 1996; pp. 215–233.

55. Diabate, M.; Munive, A.; De Faria, S.M.; Ba, A.; Dreyfus, B.; Galiana, A. Occurrence of nodulation in unexplored leguminous trees native to the West African tropical rainforest and inoculation response of native species useful in reforestation. *New Phytol.* **2005**, *166*, 231–239. [CrossRef] [PubMed]

56. Babale, A. The Interplay of Habitat and Seed Size on the Shift in Species Composition in a Fragmented Afromontane Forest Landscape: Implications for the Management of Forest Restoration. Ph.D. Thesis, University of Canterbury, Christchurch, New Zealand, 2014.

57. Ndah, N.R.; Andrew, E.E.; Bechem, E. Species composition, diversity and distribution in a disturbed Takamanda Rainforest, South West, Cameroon. *Afr. J. Plant Sci.* **2013**, *7*, 577–585.

58. Bässler, C.; Cadotte, M.W.; Beudert, B.; Heibl, C.; Blaschke, M.; Bradtka, J.H.; Langbehn, T.; Werth, S.; Müller, J. Contrasting patterns of lichen functional diversity and species richness across an elevation gradient. *Ecography* **2016**, *39*, 689–698. [CrossRef]

59. Murphy, S.J.; Audino, L.D.; Whitacre, J.; Eck, J.L.; Wenzel, J.W.; Queenborough, S.A.; Comita, L.S. Species associations structured by environment and land-use history promote beta-diversity in a temperate forest. *Ecology* **2015**, *96*, 705–715. [CrossRef]

60. Comita, L.S.; Engelbrecht, B.M. Seasonal and spatial variation in water availability drive habitat associations in a tropical forest. *Ecology* **2009**, *90*, 2755–2765. [CrossRef] [PubMed]

61. Scalley, T.H.; Crowl, T.A.; Thompson, J. Tree species distributions in relation to stream distance in a mid-montane wet forest, Puerto Rico. *Caribb. J. Sci.* **2009**, *45*, 52–64. [CrossRef]

62. Koffi, K.; Hardy, O.; Doumenge, C.; Cruaud, C.; Heuertz, M. Diversity gradients and phylogeographic patterns in Santiria trimera (Burseraceae), a widespread African tree typical of mature rainforests. *Am. J. Bot.* **2011**, *98*, 254–264. [CrossRef] [PubMed]

63. Stone, R.D. The species-rich, paleotropical genus Memecylon (Melastomataceae): Molecular phylogenetics and revised infrageneric classification of the African species. *Taxon* **2014**, *63*, 539–561. [CrossRef]
64. Brailsford, L.E. Evidence for Genetic Decline within Afromontane Forest Fragments on the Mambilla Plateau, Nigeria. Master's Thesis, University of Canterbury, Christchurch, New Zealand, 2018.

Review

The Role of Climate and Topography in Shaping the Diversity of Plant Communities in Cabo Verde Islands

Carlos Neto [1], José Carlos Costa [2], Albano Figueiredo [3], Jorge Capelo [2], Isildo Gomes [4], Sónia Vitória [5], José Maria Semedo [5], António Lopes [1], Herculano Dinis [6], Ezequiel Correia [1], Maria Cristina Duarte [7] and Maria M. Romeiras [2,7,*]

1. Centre for Geographical Studies, Institute of Geography and Spatial Planning (IGOT), Universidade de Lisboa, Rua Branca Edmée Marques, 1600-276 Lisboa, Portugal; cneto@campus.ul.pt (C.N.); antonio.lopes@campus.ul.pt (A.L.); ezequielc@campus.ul.pt (E.C.)
2. Linking Landscape, Environment, Agriculture and Food (LEAF), Instituto Superior de Agronomia (ISA), Universidade de Lisboa, 1349-017 Lisboa, Portugal; jccosta@isa.ulisboa.pt (J.C.C.); jorge.capelo@iniav.pt (J.C.)
3. Centre of Studies in Geography and Spatial Planning (CEGOT), Department Geography and Tourism, University of Coimbra, Colégio de São Jerónimo, 3004-530 Coimbra, Portugal; geofig@fl.uc.pt
4. Instituto Nacional de Investigação e Desenvolvimento Agrário (INIDA), Santiago, São Lourenço dos Orgãos CP 84, Cape Verde; isildo.gomes@inida.gov.cv
5. Campus do Palmarejo, Faculdade de Ciências e Tecnologia, Universidade de Cabo Verde, CP 279, Praia, Santiago CP 279, Cape Verde; sonia.silva@docente.unicv.edu.cv (S.V.); jmsemedo@cvtelecom.cv (J.M.S.)
6. Direção Nacional do Ambiente (DNA-CV) & Associação Projecto Vitó, CP 47, Xaguate, S. Filipe, Ilha do Fogo CP47, Cape Verde; pnfogo.segecol@gmail.com
7. Centre for Ecology, Evolution and Environmental Changes (cE3c), Faculdade de Ciências, Universidade de Lisboa, 1749-017 Lisboa, Portugal; mcduarte@fc.ul.pt
* Correspondence: mmromeiras@isa.ulisboa.pt

Received: 19 January 2020; Accepted: 15 February 2020; Published: 19 February 2020

Abstract: The flora and vegetation of the archipelago of Cabo Verde is dominated by Macaronesian, Mediterranean, and particularly by African tropical elements, resulting from its southernmost location, when compared to the other islands of the Macaronesia (i.e., Azores, Madeira, Selvagens, and Canary Islands). Very likely, such a geographical position entailed higher susceptibility to extreme climatic fluctuations, namely those associated with the West African Monsoon oscillations. These fluctuations led to a continuous aridification, which is a clear trend shown by most recent studies based on continental shelf cores. Promoting important environmental shifts, such climatic fluctuations are accepted as determinant to explain the current spatial distribution patterns of taxa, as well as the composition of the plant communities. In this paper, we present a comprehensive characterization of the main plant communities in Cabo Verde, and we discuss the role of the climatic and topoclimatic diversity in shaping the vegetation composition and distribution of this archipelago. Our study reveals a strong variation in the diversity of plant communities across elevation gradients and distinct patterns of richness among plant communities. Moreover, we present an overview of the biogeographical relationships of the Cabo Verde flora and vegetation with the other Macaronesian Islands and northwestern Africa. We discuss how the distribution of plant communities and genetic patterns found among most of the endemic lineages can be related to Africa's ongoing aridification, exploring the impacts of a process that marks northern Africa from the Late Miocene until the present.

Keywords: vegetation; aridification in NW Africa; Macaronesian islands; distribution patterns; West African Monsoon (WAM); vascular flora

---

1. Introduction

The study of deep-sea cores from the West African coast conducted in recent decades, in parallel with studies of the paleolakes that characterized the wet periods of the present Sahara desert (green Sahara Periods, or GSPs), and sedimentary structures associated with ancient rivers brought a new understanding of the history of West and North African vegetation [1–4]. Results from pollen sequences, which were used as proxies for flora and vegetation spatial patterns, were combined with data concerning dust carried from the Sahara desert, supporting the inference of paleoclimatic conditions, namely the influence of atmospheric circulation systems such as the West African Monsoon (WAM) and trade winds [5], and with changes in hydrology of the Sahara desert [6]. Such results are of critical importance to set the origin of Cabo Verde's flora and vegetation and clarify their relationships with the remaining flora and vegetation of northwestern Africa and the Mediterranean Basin.

The Atlantic archipelagos originated from mid-Tertiary volcanic cycles, spreading from 40° to 15° N along the coasts of Europe and Africa, are known as Macaronesia, or "the Fortunate Islands": Azores, Madeira, Selvagens Islands, Canary Islands, and Cabo Verde (Figure 1). The almost linear geographical disposition of the archipelagos, and their approximately parallel position in relation to the African and European mainland, creating a chain of islands, are paramount to hypotheses explaining "Macaronesian" flora and vegetation features, namely colonization, dispersion, and speciation events [7].

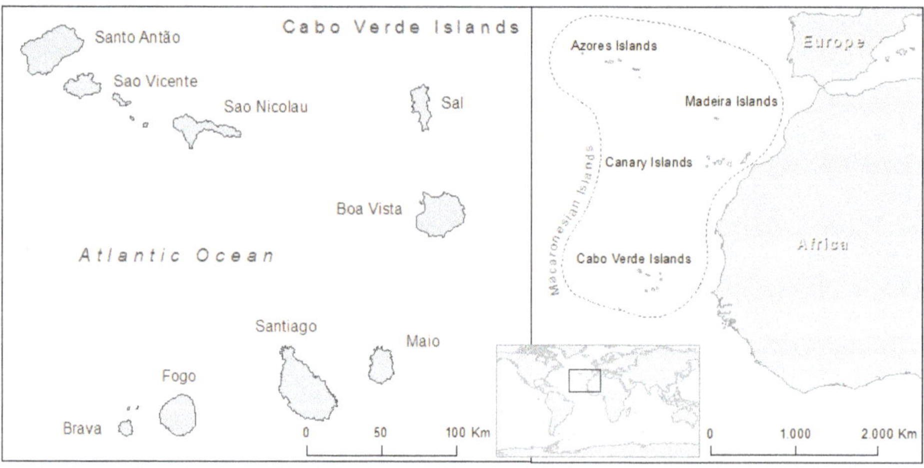

Figure 1. Cabo Verde location, with details of the Macaronesian Region.

In the continental areas surrounding Tethys, several violent disruptive environmental events took place during the mid and late Tertiary. The orogenic events associated with the alpine tectonic cycle and the onset of a summer-dry Mediterranean climate led to dramatic changes, extinctions, and novel evolutionary pressures on plant species and vegetation of those areas [8]. In this context, a global opportunity for the establishment of new migrant floras appeared, either from central Eurasia (neomediterranean) or high northern latitudes (Artho-Tertiary deciduous flora) [8]. Oceanic islands were not significantly affected by such complex phenomena during the Tertiary period, thus retaining part of the paleo-subtropical vegetation in relatively high latitudes. However, spatial changes of distribution took place, very likely contributing to the definition of low-saturated habitats, which is a feature also triggered by volcanic events, promoting windows for colonization [9,10] by creating suitable conditions for new founders [11] that very often went through a process of speciation by adaptive radiation [12]. In fact, biogeographical relationships, mostly with the Mediterranean and African mainland areas, were conditioned by long-range dispersal events.

Meanwhile, extensive speciation phenomena were taking place, which were rooted on both the ancestors from the Tertiary elements and the later newcomers from the continental areas [12]. Traditionally, the native vegetation and high prevalence of endemic elements in these islands were interpreted as relicts [13] of the subtropical Tertiary vegetation, the 'Geoflora vegetation', around the archaic Tethys basin (coarsely the Mediterranean Sea, spreading eastwards to include the present Black and Aral seas, reaching the Indian Ocean). This interpretation is challenged by increasing molecular and phylogeographical evidences [14,15] supporting the derived neoendemic synapomorphic character. This is the case of woody endemics in taxonomical groups that normally present an herbaceous habit in continental areas, which is one of the most striking attributes in the floras of the Atlantic islands [14]. In fact, the woody habit is significant among the endemics, since several genera, subgenera, or sections underwent speciation in one or in several archipelagos, namely the Asteraceae (e.g., *Asteriscus, Pericallis, Sonchus, Tolpis*), Boraginaceae (*Echium*), Campanulaceae (*Musschia* in Madeira, *Azorina* in the Azores, *Canarina* in the Canary Islands, and *Campanula* in Cabo Verde), Scrophulariaceae (*Isoplexis*) and others such as *Sideritis, Plantago,* and *Aeonium* (respectively, Lamiaceae, Plantaginaceae, and Crassulaceae). These plants exhibit consistent habit and structure features, namely being rosulate, monocarpic, or candelabra-shaped, exhibiting the so-called "island-woodiness" effect [16]. Other important groups encompassing Cabo Verde vegetation are paleo-Mediterranean thermophilous sclerophylls, such as spurges *Euphorbia* (sect. *Aphyllis*) and *Sideroxylon*. The latter lineage has circum-Mediterranean–African–Arabic affinities (Rand Flora) and reached the islands in several cycles of colonization. Rand Flora are floristic elements of taxonomical non-sister clades that are consistently found with a correlated distribution area around the peripheral borders of the whole African continent, including areas surrounding the Mediterranean and Red Seas, with no obvious relationship with inland African continental flora [17]. Such context explains why Cabo Verde's flora combines significant Macaronesian radiations (e.g., *Aeonium, Echium, Tolpis,* and *Sonchus*) with continental African elements (e.g., *Acacia, Dichrostachys, Ficus, Ziziphus,* and *Andopogoneae* grasses). In short, Cabo Verde's flora can be coarsely grouped into five types: (1) Paleoendemic paleosubtropical forest flora of tethysian origin; (2) Neoendemic Macaronesian flora with island woodiness physiognomy; (3) Paleo-Mediterranean xeric to semi-desert, sclerophyllous, or succulent flora; (4) Continental African flora; and (5) Flora introduced since the human colonization of these islands, more than 500 years ago.

2. Cabo Verde Islands: Origin and Climate

The Cabo Verde archipelago is composed of 10 main islands and several islets (Figure 1). The islands account for 4033 km^2 of total land area and are grouped into two main sets according to prevailing NE winds: The Windward islands—Santo Antão, São Vicente, Santa Luzia (the only uninhabited main island), São Nicolau, Sal, and Boavista—and Leeward islands: Maio, Santiago, Fogo, and Brava.

The genesis of the Cabo Verde archipelago is associated with intravolcanic plaque processes. According to some authors, who defend its origin from mantle plumes (hotspot) [18], there is a submerged crest connection between the Cabo Verde and Canary archipelagos, which also have a similar nature of volcanic episodes.

Hotspot activity started around 19 to 22 Ma, resulting in a large crustal uplift zone (Cabo Verde Swell) where the Cabo Verde islands are placed [19], and volcanic activity remains until present days. The topography is generally very rugged, associated with high massifs of volcanic origin, with well-preserved volcanic apparatuses, namely cones, craters, boilers (e.g., Chã das Caldeiras in Fogo and Cova in Santo Antão), and deep valleys. The island of Fogo reaches the highest elevation at 2829 m, followed by Santo Antão, Santiago, and São Nicolau with 1979 m, 1392 m, and 1304 m, respectively [20].

Due to its northern position relative to the oscillation zone of the ITCZ (Intertropical Convergence Zone), Cabo Verde has a dry tropical climate with two well-marked seasons (humid and dry) conditioning the distribution of its flora and vegetation. However, the island topography contributes to significant spatial variations with altitude and exposure to prevailing winds, leading to contrasting weather conditions. South blowing wind during the few ITCZ passages north of the archipelago brings tropical humid air masses and heavy rainfall associated to the West African Monsoon (WAM); in contrast, northeast winds, mainly in the dry season, carry hot and dry air masses (Saharan): The Harmattan season [21].

For most of the year, and almost exclusively between December and June, Cabo Verde is under the influence of the eastern sector of the Azorean subtropical high-pressure cell, which is the origin of the boreal trade winds. The oceanic course of the air mass and the vertical structure is marked by a clear thermal inversion, whose base is on average between 380 and 850 m, and the top at about 1400 m [22], favors the formation of stratiform cloud banks. The persistence of cloud banks, which is prompted by trade wind inversion on the north and northeast slopes on islands of wider altitudinal ranges, smoothens the characteristic dryness of the archipelago.

The orographic convection that occurs when the flow reaches the islands with the most vigorous reliefs reinforces the cloud thickness, which dissipates rapidly leeward due to the foehn effect, resulting in a clear dissymmetry between the northern and southern slopes. Windward areas experience a significant number of foggy days, not only at higher altitudes (e.g., Santiago Island, Serra da Malagueta, and Pico da Antónia: 1000–1300 m, about 200 days/year on average), but also in lower areas (e.g., S. Jorge dos Orgãos: 320 m, about 180 days/year). On leeward areas, there are significantly fewer foggy days (e.g., Curralinho: 950 m, 100 days/year; Assomada plateau: 550 m, 15 days/year) [21].

The high frequency of clouds, besides reducing the amount of solar radiation reaching the surface, considerably increases the water availability in these areas through the interception of cloud droplets and the increase of condensation nuclei. Since the 1960s, various cloud water-harvesting experiments have been conducted on the highest sectors of Santo Antão, São Vicente, São Nicolau, Santiago, Fogo, and the Brava islands. The results show that the horizontal precipitation far outweighs vertical precipitation, more than doubling it in some cases [23], which is a volume similar to that recorded in other Macaronesian Islands [24,25]. Although the seasonal rhythm of horizontal precipitation is similar to that of vertical precipitation, with the highest values during the rainy season, its ecological significance is greater between November and March. During the dry season, as much as 100 mm were monthly recorded at Serra da Malagueta [26], between 40 and 100 mm at Pero Dias (Santo Antão) and 176 to 1029 mm at Campo das Fontes (Brava) [27].

According to the Worldwide Bioclimatic Classification system of Rivas-Martínez et al. [28], Cabo Verde's bioclimate ranges from tropical hyperdesertic to pluviseasonal, upper infra to low supratropical, and upper ultrahyperarid to upper dry, and it is occasionally subhumid (Supplementary Materials, Table S1).

3. Plant Communities of Cabo Verde

Spatial differences in the amount of precipitation, structured by topographic determinants (altitude and aspect), become evident when comparing the flatter eastern islands (i.e., Sal, Boavista, Maio), where the climate is arid, with the most mountainous islands (Santo Antão, São Nicolau, Fogo, Santiago, Brava), where a semiarid and dry bioclimatic pattern in the south-facing slopes, dominated by African floristic elements of savannoid or predesert character, are replaced by more humid conditions on the north/northeast-facing slopes, creating a significant climatic asymmetry [28,29]. To find a reasonably effective model to understand Cabo Verde's flora and vegetation chorology, we must add the influence of successive volcanic eruptions to the atmospheric circulation model and spatial distribution of topoclimates [30], as well as the particularly devastating anthropic impact of 500 years of human colonization [31–33]. The high level of landscape disturbance on the islands, which certainly led to the extinction of some plant species, such as *Stachytarpheta fallax* or *Habenaria petromedusa*, the two formally reported extinct species [34], hinders our capacity to provide a model for the pristine vegetation. Therefore, an approach must rely on data available for other Macaronesian archipelagos and consider models of distribution of African flora and vegetation that are closest to the ones of Cabo Verde, along with historical records from navigators and/or naturalists [33,35].

Based on the aforementioned contingencies, we discuss a distribution model of the vegetation of the Cabo Verde Islands, supported by published [28,29,36] and unpublished work developed by the authors during the last decades. It must be noted that annual and ruderal communities were excluded from this study, as they are mainly dominated by exotic plants that have become naturalized in Cabo Verde. Some exotic plants are invading and dominating natural communities, thus constituting one of the most serious threats to plant communities, which are nowadays very strongly affected by human actions. According to the Cabo Verde Red List [34], most of the strong anthropic disturbances were recorded between 400 and 1200 m in mountainous islands (Santo Antão, São Nicolau, Fogo, Santiago, Brava), whereas tourism has threatened the lowland coastal plant communities, which occur in Eastern Islands of Sal and Boavista [34]. Moreover, natural disasters, specifically recent volcanic events (in 2014), have had an impact on plant communities that occur above 1600 m on Fogo Island.

In this section, we present a characterization of the main plant communities in Cabo Verde, which still constitute the dominant elements of the native vegetation of these islands [i.e., 3.1. *Ficus* and *Sideroxylon* woodlands (Figure 2A); 3.2. *Acacia* savannas (Figure 2B); 3.3. Other arborescent communities (Figure 2C,D); 3.4. Shrub vegetation (Figure 2E,F); 3.5. Grasslands (Figure 2G); 3.6. Halophytic and hydrophytic communities (Figure 2H); and 3.7. Chasmophytic communities (Figure 2I)]. Nomenclature follows Rivas-Martínez et al. [28] and POWO—The Plants of the World Online portal [37].

3.1. Ficus and Sideroxylon Woodlands

Speaking of forests in Cabo Verde today may seem somewhat speculative, as only a few native tree species occur, and especially in areas of higher humidity and/or less accessibility. The restriction of plant communities dominated by native trees is linked to the fact that in general, endemics are more vulnerable to the browsing of introduced herbivores (e.g., goats), as shown by Cubas et al. [38] for Tenerife. In addition, the harvesting of firewood for domestic use and human-induced fires are particularly destructive to the forest areas of island ecosystems [39].

In mountainous islands, the possible past occurrence of arboreal vegetation is based on the current existence of several native taxa that are capable of phanerophytic structure (e.g., *Ficus sur* and *F. sycomorus* (Figure 2A), *Dichrostachys cinerea* subsp. *platycarpa*, and *Sideroxylon marginatum*), which could form open forests or woodlands. The model considers suitable conditions for three types of woodlands, two of which are clearly dominated by tropical fig trees (*F. sur* and *F. sycomorus*) and one corresponding to open forests of *Sideroxylon marginatum* (Table S2), which are currently restricted to very small patches.

In areas with the highest vertical and horizontal precipitation values, the potential forests (woodlands) are dominated by *F. sur* at higher altitudes and by *F. sycomorus* at lower altitudes. In both cases, the phenological regime of these fig trees is similar to that of an evergreen forest, which is an outstanding feature when compared to the deciduous response pattern to the dry season found in continental Africa [40]. Outside the fog belt, sycamore fig trees (*F. sycomorus*) occur at valley bottoms, where floristic elements from arid and semi-arid environments are present, such as *Cocculus pendulus* and *Ziziphus mauritiana*, as well as tropophytic savannoid elements in an edaphic hygrophilic position according to a recurring tropical dry worldwide model [41], representing a process of isolation of high water-demanding flora and vegetation [29]. Currently, there are only a few specimens of isolated fig trees, which are rarely larger than 30 cm in diameter. These tropophytic and phreatophytic figs were first observed by Diogo Gomes (ca. 1420–1500), who was a Portuguese sailor that provided the earliest explicit reference to the large number of highly productive fruiting fig trees on the island of Santiago [33].

Figure 2. Representative species of some of the main plant communities in Cabo Verde. Woody elements present in woodlands and savannas at lowlands: *Ficus sycomorus* (**A**), *Acacia caboverdeana* (**B**), and in mountainous slopes, *Dracaena caboverdeana* (**C**). Riparian streams, near the coast: *Phoenix atlantica* (**D**). Shrub communities: *Asteriscus vogelii* and *Artemisia gorgonum* (**E**) and *Euphorbia tuckeyana* (**F**). Perennial grasslands at high altitudes: *Hyparrhenia caboverdeana* (**G**). Coastal halophytic communities: *Tetraena vicentina* (**H**). Upland chasmophytic communities: *Umbilicus schmidtii* (**I**). Photos by M. Cristina Duarte (**A–D**) and Maria Romeiras (**E–I**).

Another evergreen woodland described for Cabo Verde [28] is dominated by the endemic marmulano of Cabo Verde (*S. marginatum*), which is a small tree that, similar to the fig trees, is found only in the cracks of rocky walls, rarely with significant trunk diameters. It is a paleosubtropical paleoendemic element, which occurs in the driest parts of the fog-influenced areas, with an identical position to that of the communities of *S. mirmulans* in Madeira or the Canarian microforest of *S. canariense*. It is an evergreen woodland that includes many deciduous nanophanerophytic elements, many of them Cabo Verde endemics [e.g., *Aeonium gorgoneum*, *Echium stenosiphon*, *Euphorbia tuckeyana*,

Lavandula rotundifolia, *Lotus jacobaeus*, *Launaea picridioides*, and *Lobularia fruticosa*, among others]. Although present in almost all the islands, its abundance is higher in islands with a more significant fog belt (namely, Santo Antão, São Nicolau, Santiago, and Fogo), where it extends beyond this fog belt [28]. Such a distribution model for *S. marginatum* is not only congruent with the anthropic destruction of a large part of the population [38], but also with the ecological constraints that prevent it from colonizing drier habitats: this limitation does not apply to shrubs present in the community, since their larger ecological plasticity allows them to develop in more xeric habitats.

This model is consistent with the latest genetic analyses regarding the divergence between taxa usually associated with Rand Flora (Afro-Mediterranean phytogeographic pattern that evolutionarily relates floras of disjunct regions such as Macaronesia, Northwest and South Africa, and Southern Arabia, among others) [42–44]. Older divergences are identified in sub-humid affinity taxa [e.g., *Sideroxylon*—17.5 Ma (10–26.2 Ma)], and those with more affinities with arid territories diverge later [e.g., *Campylanthus*—7.5 Ma (3.1–14.2 Ma)] [17]. The work of Pokorny et al. [17] demonstrates that the aridification of northwestern Africa, despite successive interruptions by GSPs, caused divergences in different periods of time for different taxa, depending on their ecological requirements, which is a process that still marks current spatial distribution patterns.

3.2. Acacia Savannas

Under arid, semi-arid, and dry conditions, vegetation becomes dominated by tropical *Acacia* savannas (*Acacia caboverdeana*, Figure 2B), which are an African element in the vegetation of Cabo Verde that potentially covers most of the islands. This species, recently described as new to this archipelago by Rivas-Martínez et al. [28], was previously identified as *Faidherbia albida*. This species occurs in much of Africa in forest galleries along rivers and is a frequent floristic element in open deciduous rainforests and, mainly, savannas [45], requiring a long dry season and permanent access to ground water. Under natural conditions, *F. albida* is found close to temporary rivers and in gullies and ravines [46] or rocky areas, similar to *A. caboverdeana* in Cabo Verde, extending its roots to a depth of 8 m [47]. *Faidherbia albida* has been greatly spread by man due to its utility for livestock feeding and for undercover agriculture because of its deciduousness in the wet season [47,48]. In Cabo Verde, the potential area of *A. caboverdeana* is occupied by plantations of *Prosopis juliflora*, and only sporadically, as in Ribeira da Barca (Santiago), is it possible to find a small sample of what the pristine savannas of *Acacia* in Cabo Verde might have been. Chevalier [49] presents some photos of *A. caboverdeana* near the city of Praia (on sandy soil near the sea) with logs larger than 30 cm

These tropophytic Afrotropical savannas of *A. caboverdeana* (Table S2) are present in all islands, and they are accompanied by other micro and mesophanerophytic elements, such as *Dichrostachys cinerea* subsp. *platycarpa* and *Ziziphus mauritiana* and, less frequently, in plant communities dominated by endemic shrubs. In the understory, savannas frequently present a graminoid matrix of tropical continental African flora.

3.3. Other Arborescent Communities

Formerly relevant in structuring the woody communities in the Cabo Verde archipelago are the endemic elements *Dracaena caboverdeana* (Figure 2C) and *Phoenix atlantica* (Figure 2D), and the native non-endemic *Tamarix senegalensis*.

The symbolic dragon tree *Dracaena caboverdeana* only occurs in small patches in Santo Antão and São Nicolau [34]. Palm groves of the remarkable endangered *Phoenix atlantica* can still be seen in the eastern and southern islands of Cabo Verde, growing on riparian streams, mostly near the coast, with temporary hydromorphy. *Cocculus pendulus* and *Ziziphus mauritiana* are some of the few species present in these restricted communities.

Present in most islands of the archipelago, galleries of tamarisk grow by temporary watercourses and torrents with generally scarce intermittent flow. These communities are similar to the riparian thicket savannas with *Tamarix* spp. of West Africa (Sahara, Mauritania, and Senegal) [50].

3.4. Shrub Vegetation

The shrub formations of Cabo Verde (Figure 2E,F) are mainly found in zones of medium to high elevations above 400 m, and the levels of endemism increase with elevation, as proposed by Steinbauer et al. [51]. The shrub vegetation is dominated by endemic species, showing a comparable pattern to Canary Islands [52,53] with a clear spatial pattern, with hotspots of endemic rarity found at high elevations and in specific habitats, namely in northeast-facing cliffs, of the Natural Parks of Moroços, Cova, and Ribeira da Torre (Santo Antão); Monte Gordo (São Nicolau); Serra da Malagueta (Santiago); Monte Verde (São Vicente); and Chã das Caldeiras (Fogo Island) [54].

Among the shrub vegetation, the endemic *Euphorbia tuckeyana* (Figure 2F) define a wide diversity of plant communities (Table S2) depending on each island's endemic species and/or taxa of restricted distribution and on certain specific ecological characteristics. These communities are rich in endemic species, most of them with either a shrubby or a subshrubby habit (e.g., *Aeonium gorgoneum, Artemisia gorgonum, Campylanthus glaber, Conyza feae, Echium hypertropicum, E. vulcanorum, Euphorbia tuckeyana, Globularia amygdalifolia, Lavandula rotundifolia,* and *Periploca chevalieri*).

As stated above, most of these communities occur on NE slopes above 400 m, where higher values of precipitation occur. The presence of endemic shrub formations at lower altitudes and drier areas is associated with situations of edaphic compensation, namely block accumulation at the base of basaltic rock walls. A similar model was proposed by Esler et al. [41] for the deserts of southern Africa, where shrub and succulent vegetation becomes dominant on stony soils that receive less than 200 mm of rainfall/year.

3.5. Grasslands

In xeric environments, woody elements are rare, and communities are dominated by perennial or short-lived herbaceous grasses such as *Bothriochloa bladhii, Enneapogon desvauxii,* and *Tetrapogon cenchriformis*. In mountainous islands, these communities are enriched by *Andropogon gayanus, Chloris pycnothrix, Heteropogon melanocarpus, Hyparrhenia caboverdeana* (Figure 2G), *Setaria parviflora,* and *Rottboellia cochinchinensis,* among others. As a whole, such herbaceous communities are physiognomically and structurally close to the dry tropical West African graminoid vegetation, sharing numerous taxa with it [28].

After rainy periods, ephemeral grasslands composed of annual grasses, such as the endemics *Aristida cardosoi* and *Brachiaria lata* subsp. *caboverdeana,* and broad-leaved herbs, such as *Cleome viscosa, C. brachycarpa* and *Heliotropium crispum,* flourish in the lowlands of all the islands.

3.6. Halophytic and Hydrophytic Communities

Halophytic communities are common in Cabo Verde, but they are particularly important in the Eastern islands of Sal, Boavista, and Maio, as well as in S. Vicente. In coastal sand dunes, the xerophilous communities are dominated by perennial rhizomatous species such as the graminoids *Sporobolus spicatus* and *Cyperus crassipes,* two widespread African species, or by the dwarf shrubs *Lotus brunneri, Polycarpaea caboverdeana, Pulicaria diffusa,* and *Tetraena vicentina* (Figure 2H).

The particular topographic and hydrographic features of the eastern islands favor the occurrence of saltwater marshes. These permanent succulent halophilous communities grow on coastal sandy soils, are temporarily or occasionally flooded by sea tides, and are dominated by the succulent shrub *Arthrocaulon franzii* and the small forb *Cressa salina*. Similar communities are present in the western coasts of the African continent, mainly from Morocco to Senegal [51]. Intermingled with these woody marshes, small herbaceous communities of succulents occur, including *Sesuvium portulacastrum* subsp. *persoonii, S. sesuvioides,* and *Blutaparon vermiculare*.

3.7. Chasmophytic Communities

Humid and shady basaltic walls, permanently or temporarily water-flushed, support perennial communities of chasmophytic species such as the tropical ferns *Adiantum capillus-veneris* subsp. *trifidum*, *Hypodematium crenatum*, and *Pteris vittata*.

In mountainous islands, perennial rupicolous communities grow on basaltic, mafic ultramafic rocky walls and cliffs that are wetted only during the rainy season. Characteristic species vary with the island and comprise the pteridophytes *Hemionitis acrostica* and *Cosentinia vellea* and shrub or subshrub chasmophytes, including a wide diversity of endemic taxa belonging to genera such as *Campylanthus*, *Kickxia*, *Launaea*, *Sonchus*, and *Umbilicus* (Figure 2I). Ferns such as *Adiantum incisum* and *A. philippense*, as well as some cosmopolite rupicolous species are present in nitrogen-rich habitats in urban or rural areas. These communities develop under less humid conditions than those previously listed.

4. Aridification of Northern Africa and Its Role in Shaping the Vegetation of Cabo Verde

Explanation of the current features of the archipelago's flora and vegetation requires arguments concerning the climatic changes that have occurred in Africa since the Miocene. The early Miocene (23–16 Ma) was globally hot and humid, with rainforests extending from northern Africa to nearly southern Africa [17]. The Middle Miocene (16–11.6 Ma) marks the beginning of the change toward aridity, although this aridification began in the SW (Namib Desert, Congo Basin) 17–16 Ma earlier (lower Mid-Miocene), and in North Africa, the first evidence of Saharan aridification only appears in the Late Miocene (11.6–5.3 Ma) [55]. However, this aridification trend is marked across North and Northwest Africa by strong aridity waves interrupted by humidity episodes known as GSPs (green Sahara Periods) [56,57]. More than 230 GSPs since 8 Ma have been documented [58], and their importance for vegetation distribution models in North and Northwest Africa cannot be discarded [2,17,56,58–62]. The alternate episodes of humidity and aridity of the aridification trend of northern Africa since the Miocene are the keystone of the irradiation phenomena of North African flora that, depending on their ecological valence, diverged into isolated and genetically differentiated populations following a similar model, but these are temporally different due to their ecology (the most humidity demanding elements present older divergences) [17,63,64]. In the Quaternary, two of these GSPs, respectively Eemian (130–115 Ky corresponding to Marine Isotope Stage 5e) and the early Holocene (11–5 ka), are relatively well documented [2,54,55,59,65,66] and are of great importance for the evolution of northern and northwestern African vegetation (including the Cabo Verde archipelago). During GSPs, WAM penetrates further north, and higher precipitation than today generated a series of inland Sahara lakes and permanent rivers [3,4]. During the Holocene GSP, the vegetation is characterized by an advance of wooded savannas northwards, occupying what is now the arid and hyperarid Sahara Desert [17]. The currently accepted general model refers to a progressive development of forests and wooded savannas during interglacial periods as a result of the intensification and latitude rise of WAM (with increased precipitation) and a decrease in vegetation cover during glaciations with a greater influence of continental trade winds. In the continental shelf cores, dusts of terrestrial origin present a minimum concentration during the interglacial periods (higher vegetation cover) and the opposite during the glacial periods. In Cabo Verde Islands, such an oscillation is recorded in the alternation of calcrete (hot and arid conditions), sand dunes (dry and cold conditions), and paleosols (wetter periods) [67,68].

In general, the endemic shrub species have a strong ecological resilience, resulting from the filtering effect of recurrent dry periods that mark the ongoing African aridification since the terminal Miocene [63]. This aridification must have led to mass extinction phenomena, and only taxa of greater plasticity managed to survive. We can assume that episodes of dryness played a more important role in Cabo Verde than in the other islands of Macaronesia. Such an assumption is based on the fact that arid and semi-arid conditions have been expanding in altitude in southernmost Macaronesian archipelagos, in a model that inevitably leads to the restriction/disappearance of the higher humid and colder zones. Such an idea is also supported by analyses of deep-sea cores from the African west coast, some of which fall between Cabo Verde and the Canary Islands. Analyses of pollen [69,70] and of Saharan dust (among others) from marine sedimentary series show that the recurrent periods of intensification of WAM affected the Cabo Verde archipelago much more than the rest of Macaronesia due to its geographical position. Such oscillations between arid/cooler and warmer/wet phases are also recorded in sedimentary sequences in all Macaronesian islands, and Cabo Verde is no exception, despite its southernmost position (see Figueiredo [71]). The recurrence of severe aridity periods, interspersed with periods of WAM intensification, led to several repetitions of the contraction and expansion phenomenon. For instance, the genetic patterns of *E. tuckeyana*, the main constituent of Cabo Verde's shrub formations, occur from sea level to near 2000 m in Fogo Island, with great ecological plasticity. In a recent work about the phylogeography of sect. *Aphyllis* subsect. *Macaronesicae* (Euphorbiaceae), Barres et al. [72] suggest that the ancestor of subsect. *Macaronesicae* was adapted to arid or mesic habitats, and the traits associated with humid habitats were acquired later. Tenerife island (Canary Islands) is considered the origin of the group, and two colonization events are identified, one of them originating *E. tuckeyana* in Cabo Verde, which shows an ancient separation from the ancestor of the rest of the species present in the other Macaronesian archipelagos. The low genetic differentiation found within Cabo Verde populations suggests a recent colonization, although the coincident arrival of the taxon with the beginning of the aridification of north and northwest Africa and climate change very likely promoted divergence between *E. tuckeyana* of Cabo Verde and the rest of sect. *Aphyllis* (5.5 Ma) [17]. Barres et al. [72] document differentiation between the populations of the northern and southern Cabo Verde Islands, which is also reported for other lineages, such as *Globularia amygdalifolia* and *Umbilicus schmidtii* [54], *Echium* [73], and *Campanula jacobaea* [74]. Menezes et al. [75] estimate 9.56 Ma as the age of divergence between Cabo Verde's endemic *Campanula* species (*C. jacobaea* and *C. bravensis*) and their African sister species, i.e., it is related to the beginning of the aridification process of North and Northwest Africa. However, the divergence between *C. jacobaea* and *C. bravensis* is recent (Pleistocene, 0.01 Ma) [75].

The genetic patterns found in Cabo Verde for the mentioned lineages were probably shaped by a combination of climate-driven expansion/contraction, recolonization, and extinction, following the tertiary and quaternary climate fluctuations discussed above. These climatic oscillations can be the true sculptors of the distribution model of flora and vegetation in Cabo Verde, as they are in continental Africa [76]. In fact, the successive periods of contraction and expansion controlled by WAM fluctuations mark the genetic patterns and the current distribution of plant species and, therefore, of communities in Cabo Verde archipelago. We can speculate that some Macaronesian taxa reached the archipelago (for example *E. tuckeyana*) in the late Miocene, during wetter periods than the current one, followed by periods of great aridity, which are responsible for the significant contraction of the established populations. Most taxa that form the dominant shrub communities in Cabo Verde apparently have high resilience and great ecological valence (especially in relation to dryness), which is certainly related with the original lapilli system that allowed subsurface water retention or even the frequent block accumulation at the base of the numerous basaltic rock walls. The expansion of these shrub communities during favorable climatic phases must have been rapid, with available habitats and little competition, especially if occurring after volcanic events.

Moreover, the rise and fall of the sea levels also plays an important role when studying the evolutionary processes in Cabo Verde due to their influence on the available terrestrial area of the islands [77]. For instance, during the glacial periods, the terrestrial area increased, promoting high gene flow among populations across the islands, followed by isolation after the retraction of populations to the middle-upper areas of the mountains during the interglacial periods (reviewed by Romeiras et al. [78]).

Although the above-mentioned hypotheses can contribute to explain some of the patterns of diversity found in these islands, the distribution of plant communities is currently strongly conditioned by human actions, namely habitat destruction and the introduction of non-native taxa [34]. These anthropogenic pressures interact with climate change impacts and will increase threats to insular floras during the 21st century [79,80]. Over the past decades, Cabo Verde has been affected by serious and prolonged droughts [81] that have negative impacts on the distribution and abundance of the endemic plant species, which have already small population sizes and occur in fragmented habitats [34].

5. The Lack of Cloud Forests Dominated by Lauroid Taxa

A remarkable feature of the Cabo Verde shrublands is clearly associated to their composition: under similar bioclimatic conditions as in Madeira and in the Canary Islands, namely in areas of high prevalence of cloud-banks, it is possible to find some species belonging to major Macaronesian radiations (e.g., *Aeonium, Campylanthus, Echium, Limonium, Sonchus, Tolpis*) [10,54,78,82]. However, data supporting the possibility that Lauraceae reached the Cabo Verde archipelago are lacking [7,83]. A recent paleoecological study by Castilla-Beltrán et al. [84] on the vegetation of São Nicolau's highlands has shed light on these questions, showing a pre-human landscape (5900–410 cal yr BP) characterized by shrubs and trees including *Euphorbia tuckeyana, Ficus, Dracaena,* and *Tamarix*.

There is no evidence that the laurel forest found in Azores, Madeira, and the Canary Islands survived in Cabo Verde, contributing to increase the debate about the definition of the Macaronesian Region as a biogeographic unit. For instance, the endemic shrub vegetation, which is mainly found in mountain areas (see Section 3.4), is closely related to other lineages from the Canary and Madeira archipelagos. On the other hand, the native grass species occurring in the arid lowlands of the archipelago (see Section 3.5 grassland communities) share more affinities with Tropical Africa. Therefore, the position of Cabo Verde is currently interpreted from a new perspective, following recent advances with other taxonomic studies (for more details, see Freitas et al. [85]).

6. Conclusions

The most recent studies on the vegetation of Cabo Verde highlight the coexistence between Macaronesian and Mediterranean floristic elements, and African tropical continental flora. The evolutionary history of Cabo Verde's endemic lineages is still understudied, particularly when compared to the other Macaronesian archipelagos [78]. Recent studies of some Cabo Verde's plant lineages, coupling morphological and genetic analyses, revealed unexpectedly high diversity and putative undescribed species [54]. Nevertheless, divergences within the archipelago are recent (for more details, see Romeiras et al. [78]), leading us to conclude that they possibly are recent expansions. This could be associated with climatic fluctuations that affect all northern Africa since the late Miocene until present, causing repeated phenomena of expansion and contraction of taxa distribution and, very likely, extinction, associated with the West African Monsoon oscillations. Such oscillations led to more significant impacts in Cabo Verde archipelago than in the other Macaronesian Islands because of its geographical position. Thus, interpretation of current distribution patterns of Cabo Verde's vegetation, as well as of genetic patterns, is linked with increasing dryness (climate-driven extinction) intercalated with more humid periods, favoring rapid expansion phenomena. Such successions of wet and dry periods shaped the particular features of the flora and vegetation of Cabo Verde relative to the remaining Macaronesian archipelagos, namely the more considerable presence of African elements, which is clearly observed in the fog belt colonized by African continental forest species of possibly recent colonization rather than by Lauraceae.

Supplementary Materials: The following are available online at http://www.mdpi.com/1424-2818/12/2/80/s1, Table S1: Bioclimates, thermotypes, and ombrotypes present in Cabo Verde (adapted from Rivas-Martínez et al. [28]), Table S2: Plant communities present in Cabo Verde (excluding ruderal communities), bioclimatic characterization and respective syntaxonomic framework [Class (Cl.), Order (Or.), Alliance (Al.), Association (As.)] (adapted from Rivas-Martínez et al. [28]).

Author Contributions: Conceptualization, C.N.; draft preparation, C.N., M.M.R.; investigation, C.N., J.C.C., M.C.D., and M.M.R.; writing—review and editing, C.N, M.C.D., and M.M.R. Revised the final version of this manuscript C.N, J.C.C., A.F., J.C., I.G., S.V., J.M.S., A.L.; H.D.; E.C., M.C.D., and M.M.R. All authors have read and agreed to the published version of the manuscript.

Funding: This research was funded by Fundação para a Ciência e Tecnologia (FCT) and Aga Khan Development Network (AKDN) under the project CVAgrobiodiversity/333111699. Also to research units: UID/AGR/04129/2020 (LEAF) and UID/BIA/00329/2020 (cE3c), funded by Portuguese National Funds through FCT, Portugal.

Acknowledgments: The authors would like to acknowledge the supported provided by Fundação para a Ciência e Tecnologia (FCT) and Aga Khan Development Network (AKDN).

Conflicts of Interest: The authors declare no conflict of interest. The funders had no role in the design of the study; in the collection, analyses, or interpretation of data; in the writing of the manuscript, or in the decision to publish the results.

References

1. Armitage, S.J.; Bristow, C.S.; Drake, N.A. West African monsoon dynamics inferred from abrupt fluctuations of Lake Mega-Chad. *Proc. Natl. Acad. Sci. USA* **2015**, *112*, 8543–8548. [CrossRef] [PubMed]

2. Grant, K.M.; Rohling, E.J.; Westerhold, T.; Zabel, M.; Heslop, D.; Konijnendijk, T.; Lourens, L. A 3 million year index for North African humidity/aridity and the implication of potential pan-African Humid periods. *Quat. Sci. Rev.* **2017**, *171*, 100–118. [CrossRef]

3. Skonieczny, C.; Paillou, P.; Bory, A.; Bayon, G.; Biscara, L.; Crosta, X.; Eynaud, F.; Malaizé, B.; Revel, M.; Aleman, N.; et al. African humid periods triggered the reactivation of a large river system in Western Sahara. *Nat. Commun.* **2015**, *6*, 6–11. [CrossRef] [PubMed]

4. Wu, J.; Liu, Z.; Stuut, J.B.W.; Zhao, Y.; Schirone, A.; de Lange, G.J. North-African paleodrainage discharges to the central Mediterranean during the last 18,000 years: A multiproxy characterization. *Quat. Sci. Rev.* **2017**, *163*, 95–113. [CrossRef]

5. Rognon, P.; Coudé-Gaussen, G. Paleoclimates Off Northwest Africa (28°–35°N) about 18,000 yr B.P. Based on Continental Eolian Deposits. *Quat. Res.* **1996**, *46*, 118–126. [CrossRef]

6. Kröpelin, S.; Verschuren, D.; Lézine, A.-M.; Eggermont, H.; Cocquyt, C.; Francus, P.; Cazet, J.-P.; Fagot, M.; Rumes, B.; Russell, J.M.; et al. Climate-Driven Ecosystem Succession in the Sahara: The Past 6000 Years. *Science* **2008**, *320*, 765–768. [CrossRef]

7. Fernández-Palacios, J.M.; De Nascimento, L.; Otto, R.; Delgado, J.D.; García-Del-Rey, E.; Arévalo, J.R.; Whittaker, R.J. A reconstruction of Palaeo-Macaronesia, with particular reference to the long-term biogeography of the Atlantic island laurel forests. *J. Biogeogr.* **2011**, *38*, 226–246. [CrossRef]

8. Carrión, J.S.; Fernández, S.; Jiménez-Moreno, G.; Fauquette, S.; Gil-Romera, G.; González-Sampériz, P.; Finlayson, C. The historical origins of aridity and vegetation degradation in southeastern Spain. *J. Arid. Environ.* **2010**, *74*, 731–736. [CrossRef]

9. Patino, J.; Whittaker, R.J.; Borges, P.A.; Fernández-Palacios, J.M.; Ah-Peng, C.; Araújo, M.B.; Ávila, S.P.; Cardoso, P.; Cornuault, J.; de Boer, E.J.; et al. A roadmap for island biology: 50 fundamental questions after 50 years of The Theory of Island Biogeography. *J. Biogeogr.* **2017**, *44*, 963–983. [CrossRef]

10. Carine, M.A. Spatio-Temporal Relationships of the Macaronesian Endemic Flora: A Relictual Series or Window of Opportunity? *Taxon* **2005**, *54*, 895–903. [CrossRef]

11. Whittaker, R.J.; Triantis, K.A.; Ladle, R.J. A general dynamic theory of oceanic island biogeography. *J. Biogeogr.* **2008**, *35*, 977–994. [CrossRef]

12. Kim, S.-C.; McGowen, M.R.; Lubinsky, P.; Barber, J.C.; Mort, M.E.; Santos-Guerra, A. Timing and Tempo of Early and Successive Adaptive Radiations in Macaronesia. *PLoS ONE* **2008**, *3*, e2139. [CrossRef]

13. Bramwell, D. Endemism in the flora of the Canary Islands. In *Taxonomy, Phytogeography and Evolution*; Valentine, D.H., Ed.; Academic Press: Cambridge, MA, USA, 1972.

14. Böhle, U.-R.; Hilger, H.H.; Martin, W.F. Island colonization and evolution of the insular woody habit in *Echium*, L. (Boraginaceae). *Proc. Natl. Acad. Sci. USA* **1996**, *93*, 11740–11745. [CrossRef] [PubMed]

15. Kim, S.-C.; Crawford, D.J.; Francisco-Ortega, J.; Santos-Guerra, A. A common origin for woody *Sonchus* and five related genera in the Macaronesian Islands: Molecular evidence for extensive radiation. *Proc. Natl. Acad. Sci. USA* **1996**, *93*, 7743–7748. [CrossRef]

16. Lens, F.; Davin, N.; Smets, E.; del Arco, M. Insular woodiness on the Canary Islands: A remarkable case of convergent evolution. *Int. J. Plant. Sci.* **2013**, *174*, 992–1013. [CrossRef]

17. Pokorny, L.; Riina, R.; Mairal, M.; Meseguer, A.S.; Culshaw, V.; Cendoya, J.; Sanmartín, I. Living on the edge: Timing of Rand Flora disjunctions congruent with ongoing aridification in Africa. *Front. Genet.* **2015**, *6*, 1–15. [CrossRef]

18. Holmes, J.A. How the Sahara became dry. *Science* **2008**, *320*, 752–753. [CrossRef]
19. Plesner, S.; Holm, P.M.; Wilson, J.R. 40-39 Ar geochronology of Santo Antão, Cape Verde Islands. *J. Volcanol. Geoth. Res.* **2003**, *120*, 103–121. [CrossRef]
20. Victória, S. Caracterização geológica e geotécnica das unidades litológicas da Cidade da Praia, ilha de Santiago, Cabo Verde. Ph.D. Thesis, University of Coimbra, Coimbra, Portugal, 2013. Available online: http://www.portaldoconhecimento.gov.cv/handle/10961/2471 (accessed on 17 February 2020).
21. Correia, E. Contribuições para o conhecimento do clima de Cabo Verde. *Garcia de Orta Sér. Geogr.* **1996**, *15*, 81–107.
22. Soares, E. Variabilidade climática na região de Cabo Verde. Master's Thesis, University of Évora, Évora, Portugal, 2004. Available online: http://www.icterra.pt/ (accessed on 17 February 2020).
23. Furtado, F.J.R. A Captação de Água no Nevoeiro no Parque Natural de Serra Malagueta. Master's Thesis, University of Aveiro, Aveiro, Portugal, 2009. Available online: http://hdl.handle.net/10773/629 (accessed on 17 February 2020).
24. Ritter, A.; Regalado, C.M.; Aschan, G. Fog reduces transpiration in tree species of the Canarian relict heath-laurel cloud forest (Garajonay National Park, Spain). *Tree Physiol.* **2009**, *29*, 517–528. [CrossRef]
25. Prada, S.; de Sequeira, M.M.; Figueira, C.; Vasconcelos, R. Cloud water interception in the high altitude tree heath forest (*Erica arborea* L.) of Paul da Serra Massif (Madeira, Portugal). *Hydrol. Process.* **2012**, *26*, 202–212. [CrossRef]
26. Acosta Baladón, A.; Gioda, A. L'importance des précipitations occultes sous les tropiques secs. *Sécheresse* **1991**, *2*, 132–134.
27. Cunha, F.R. Problema da Captação da Água de Nevoeiro em Cabo Verde. *Garcia de Orta* **1964**, *12*, 719–754.
28. Rivas-Martínez, S.; Lousã, M.; Costa, J.C.; Duarte, M.C. Geobotanical survey of Cabo Verde Islands (West Africa). *Int. J. Geobot. Res.* **2017**, *7*, 1–103.
29. Duarte, M.C.; Rego, F.; Moreira, I. Distribution patterns of plant communities on Santiago Island, Cape Verde. *J. Veg. Sci.* **2005**, *16*, 283–292. [CrossRef]
30. Díaz-Pérez, A.; Sequeira, M.; Santos-Guerra, A.; Catalán, P. Multiple colonizations, in situ speciation, and volcanism-associated stepping-stone dispersals shaped the phylogeography of the Macaronesian red fescues (*Festuca*, L., Gramineae). *Syst. Biol.* **2008**, *57*, 732–749. [CrossRef]
31. Duarte, M.C.; Moreira, I. A vegetação de Santiago (Cabo Verde). Apontamento histórico. *Garcia de Orta Sér. Bot.* **2002**, *16*, 51–80.
32. Caujapé-Castells, J.; Tye, A.; Crawford, D.J.; Santos-Guerra, A.; Sakai, A.; Beaver, K.; Lobin, W.; Vincent Florens, F.B.; Moura, M.; Jardim, R.; et al. Conservation of oceanic island floras: Present and future global challenges. *Perspect. Plant Ecol.* **2010**, *12*, 107–129. [CrossRef]
33. Romeiras, M.M.; Duarte, M.C.; Santos-Guerra, A.; Carine, M.; Francisco-Ortega, J. Botanical exploration of the Cape Verde Islands: From the pre-Linnaean records and collections to late 18th century floristic accounts and expeditions. *Taxon* **2014**, *63*, 625–640. [CrossRef]
34. Romeiras, M.M.; Catarino, S.; Gomes, I.; Fernandes, C.; Costa, J.C.; Caujapé-Castells, J.; Duarte, M.C. IUCN Red List assessment of the Cape Verde endemic flora: Towards a Global Strategy for Plant Conservation in Macaronesia. *Bot. J. Linn. Soc.* **2016**, *180*, 413–425. [CrossRef]
35. Rico, L.; Duarte, M.C.; Romeiras, M.M.; Santos-Guerra, A.; Nepi, C.; Francisco-Ortega, J.; Joseph, D. Hooker's 1839 Cabo Verde Collections. *Curtis's Bot. Mag.* **2017**, *34*, 146–168. [CrossRef]
36. Brochmann, C.; Rustan, Ø.H.; Lobin, W.; Kilian, N. The endemic vascular plants of the Cape Verde Islands, W Africa. *Sommerfeltia* **1997**, *24*, 1–356.
37. POWO. Plants of the World Online. *Royal Botanic Gardens, Kew.* Available online: www.plantsoftheworldonline.org (accessed on 8 January 2020).
38. Cubas, J.; Martín-Esquivel, J.L.; Nogales, M.; Irl, S.D.; Hernández-Hernández, R.; López-Darias, M.; González-Mancebo, J.M. Contrasting effects of invasive rabbits on endemic plants driving vegetation change in a subtropical alpine insular environment. *Biol. Invasions* **2018**, *20*, 793–807. [CrossRef]
39. Irl, S.D.; Steinbauer, M.J.; Messinger, J.; Blume-Werry, G.; Palomares-Martínez, Á.; Beierkuhnlein, C.; Jentsch, A. Burned and devoured-introduced herbivores, fire, and the endemic flora of the high-elevation ecosystem on La Palma, Canary Islands. *Arct. Antarct. Alp. Res.* **2018**, *46*, 859–869. [CrossRef]
40. Burrows, J.; Burrows, S. *Figs of Southern and South.-Central Africa*; Umdaus Press: Hatfield, South Africa, 2003.

41. Esler, K.J.; Milton, S.; Dean, W.R.J. *Karoo Veld: Ecology and Management*; Briza Publications: Pretoria, South Africa, 2006; p. 214.

42. Andrus, N.; Trusty, J.; Santos-Guerra, A.; Jansen, R.K.; Francisco-Ortega, J. Using molecular phylogenies to test phytogeographical links between East/South Africa–Southern Arabia and the Macaronesian islands—A review, and the case of *Vierea* and *Pulicaria* section Vieraeopsis (Asteraceae). *Taxon* **2004**, *53*, 333–346. [CrossRef]

43. Sanmartin, I.; Anderson, C.L.; Alarcon, M.; Ronquist, F.; Aldasoro, J.J. Bayesian island biogeography in a continental setting: The Rand Flora case. *Biol. Lett.* **2010**, *6*, 703–707. [CrossRef]

44. Mairal, M.; Sanmartín, I.; Herrero, A.; Pokorny, L.; Vargas, P.; Aldasoro, J.; Alarcón, M. Geographic barriers and Pleistocene climate change shaped patterns of genetic variation in the Eastern Afromontane biodiversity hotspot. *Sci. Rep.* **2017**, *7*, 45749. [CrossRef]

45. Wood, P.J. The Botany and Distribution of *Faidherbia albida*. In *Faidherbia albida in the West African Semi-Arid Tropics*; Vandenbeldt, R.J., Ed.; International Centre for Research in Agroforestry: Niamey, Niger, 1992.

46. Joly, H. The Genetics of *Acacia albida* (syn. *Faidherbia albida*). In *Faidherbia albida in the West African Semi-Arid Tropics*; Vandenbeldt, R.J., Ed.; International Centre for Research in Agroforestry: Niamey, Niger, 1992.albida). In *Faidherbia albida in the West African Semi-Arid Tropics*; Vandenbeldt, R.J., Ed.; International Centre for Research in Agroforestry: Niamey, Niger, 1992; Vandenbeldt, R.J., Ed.

47. Roupsard, O.; Ferhi, A.; Granier, A.; Pallo, F.; Depommier, D.; Mallet, B.; Joly, H.I.; Dreyer, E. Reverse phenology and dry season water uptake by *Fadherbia albida* (Del.) A. Chev. in an agroforestry parkland of Sudanense Africa. *Funct. Ecol.* **1999**, *13*, 460–472. [CrossRef]

48. Kirmse, R.; Norton, B. The potential of *Acacia albida* for desertification control and increased productivity in Chad. *Biol. Conserv.* **1984**, *29*, 121–141. [CrossRef]

49. Chevalier, A. Les Iles du Cap Vert. Géographie, biogéographie, agriculture. Flore de l'archipel. *Rev. Bot. Appl. Agric. Trop.* **1935**, *15*, 733–1090. [CrossRef]

50. Schulz, E.; Abichou, A.; Adamou, A.; Ousseïni, I.; Ballouche, A. The desert in the Sahara. Transitions and boundaries. In *Palaeoecology of Africa and the Surrounding Islands*; Baumhauer, R., Runge, J., Eds.; Taylor & Francis Group: London, UK, 2009; Volume 29, pp. 63–89.

51. Steinbauer, M.J.; Field, R.; Grytnes, J.A.; Trigas, P.; Ah-Peng, C.; Attorre, F.; Beierkuhnlein, C. Topography-driven isolation, speciation and a global increase of endemism with elevation. *Glob. Ecol. Biogeogr.* **2016**, *25*, 1097–1107. [CrossRef]

52. Irl, S.D.H.; Schweiger, A.H.; Medina, F.M.; Fernández-Palacios, J.M.; Harter, D.E.; Jentsch, A.; Beierkuhnlein, C. An island view of endemic rarity—Environmental drivers and consequences for nature conservation. *Divers. Distrib.* **2017**, *23*, 1132–1142. [CrossRef]

53. Otto, R.; Whittaker, R.J.; von Gaisberg, M.; Stierstorfer, C.; Naranjo-Cigala, A.; Steinbauer, M.J.; Fernández-Palacios, J.M. Transferring and implementing the general dynamic model of oceanic island biogeography at the scale of island fragments: The roles of geological age and topography in plant diversification in the Canaries. *J. Biogeogr.* **2016**, *43*, 911–922. [CrossRef]

54. Romeiras, M.M.; Monteiro, F.; Duarte, M.C.; Schaefer, H.; Carine, M. Patterns of genetic diversity in three plant lineages endemic to the Cape Verde Islands. *AoB PLANTS* **2015**, *7*, 1–11. [CrossRef]

55. Senut, B.; Pickford, M.; Ségalen, L. Neogene desertification of Africa. *C. R. Geosci.* **2009**, *341*, 591–602. [CrossRef]

56. Adkins, J.; DeMenocal, P.; Eshel, G. The "African humid period" and the record of marine upwelling from excess 230Th in Ocean Drilling Program Hole 658C. *Paleoceanography* **2006**, *21*, 1–14. [CrossRef]

57. Pausata, F.S.R.; Messori, G.; Zhang, Q. Impacts of dust reduction on the northward expansion of the African monsoon during the Green Sahara period. *Earth Planet. Sci. Lett.* **2016**, *434*, 298–307. [CrossRef]

58. Larrasoaña, J.C.; Roberts, A.P.; Rohling, E.J. Dynamics of Green Sahara Periods and Their Role in Hominin Evolution. *PLoS ONE* **2013**, *8*, e76514. [CrossRef]

59. Jung, S.J.A.; Davies, G.R.; Ganssen, G.M.; Kroon, D. Stepwise Holocene aridification in NE Africa deduced from dust-borne radiogenic isotope records. *Earth Planet. Sci. Lett.* **2004**, *221*, 27–37. [CrossRef]

60. Kuper, R.; Kröpelin, S. Climate-Controlled Holocene Occupation in the Sahara: Motor of Africa's Evolution. *Science* **2006**, *313*, 803–807. [CrossRef]

61. Liu, Z.; Wang, Y.; Gallimore, R.; Gasse, F.; Johnson, T.; DeMenocal, P.; Adkins, J.; Notaro, M.; Prentice, I.C.; Kutzbach, J.; et al. Simulating the transient evolution and abrupt change of Northern Africa atmosphere-ocean-terrestrial ecosystem in the Holocene. *Quat. Sci. Rev.* **2007**, *26*, 1818–1837. [CrossRef]

62. Wright, D.K. Humans as Agents in the Termination of the African Humid Period. *Front. Earth Sci.* **2017**, *5*, 4. [CrossRef]

63. Mairal, M.; Pokorny, L.; Aldasoro, J.J.; Alarcõn, M.; Sanmartín, I. Ancient vicariance and climate-driven extinction continental-wide disjunctions in Africa: The case of the Rand Flora genus *Canarina* (Campanulaceae). *Mol. Ecol.* **2015**, *24*, 1335–1354. [CrossRef] [PubMed]

64. Navarro-Pérez, M.L.; Vargas, P.; Fernández-Mazuecos, M.; López, J.; Valtueña, F.J.; Ortega-Olivencia, A. Multiple windows of colonization to Macaronesia by the dispersal-unspecialized *Scrophularia* since the Late Miocene. *Perspect. Plant. Ecol.* **2015**, *17*, 263–273. [CrossRef]

65. Dupont, L.M.; Jahns, S.; Marret, F.; Ning, S. Vegetation change in equatorial West Africa: Time-slices for the last 150 ka. *Palaeogeogr. Palaeocl.* **2000**, *155*, 95–122. [CrossRef]

66. Dalibard, M.; Popescu, S.M.; Maley, J.; Baudin, F.; Melinte-Dobrinescu, M.C.; Pittet, B.; Marsset, T.; Bernard, D.; Laurence, D.; Suc, J.P. High-resolution vegetation history of West Africa during the last 145 ka. *Geobios* **2014**, *47*, 183–198. [CrossRef]

67. Zazo, C.; Goy, J.L.; Dabrio, C.J.; Soler, V.; Hillaire-Marcel, C.; Ghaleb, B.; González-Delgado, J.A.; Bardají, T.; Cabrero, A. Quaternary marine terraces on Sal Island (Cape Verde archipelago). *Quat. Sci. Rev.* **2007**, *26*, 876–893. [CrossRef]

68. Zazo, C.; Goy, J.L.; Dabrio, C.J.; Cabero, A.; Bardaji, T.; Ghaleb, B.; Soler, V. Sea level changes during the last and present interglacials in Sal Island (Cape Verde archipelago). *Glob. Planet. Chang.* **2010**, *72*, 302–317. [CrossRef]

69. Hooghiemstra, H.; Lézine, A.-M.; Leroy, S.A.G.; Dupont, L.; Marret, F. Late Quaternary palynology in marine sediments: A synthesis of the understanding of pollen distribution patterns in the NW African setting. *Quat. Int.* **2006**, *148*, 29–44. [CrossRef]

70. Dupont, L. Orbital scale vegetation change in Africa. *Quat. Sci. Rev.* **2011**, *30*, 3589–3602. [CrossRef]

71. Figueiredo, A. Assessing Climate Change Impacts on the Distribution of Flora and Vegetation at Madeira Island. Ph.D. Thesis, Universidade de Coimbra, Coimbra, Portugal, 2013.

72. Barres, L.; Galbany-Casals, M.; Hipp, A.; Molero, J.; Vilatersana, R. Phylogeography and character evolution of *Euphorbia* sect. Aphyllis subsect. Macaronesicae (Euphorbiaceae). *Taxon* **2017**, *2*, 324–342. [CrossRef]

73. Romeiras, M.M.; Paulo, O.S.; Duarte, M.C.; Pina-Martins, F.; Cotrim, M.H.; Carine, M.A.; Pais, M.S. Origin and diversification of genus *Echium* (Boraginaceae) in Cape Verde islands: A phylogenetic study based on ITS (rDNA) and cpDNA sequences. *Taxon* **2011**, *60*, 1375–1385. [CrossRef]

74. Alarcón, M.; Roquet, C.; García-Fernández, A.; Vargas, P.; Aldasoro, J.J. Phylogenetic and phylogeographic evidence for a Pleistocene disjunction between *Campanula jacobaea* (Cape Verde Islands) and *C. balfourii* (Socotra). *Mol. Phylogenet. Evol.* **2013**, *69*, 828–836. [CrossRef] [PubMed]

75. Menezes, T.; Romeiras, M.M.; de Sequeira, M.M.; Moura, M. Phylogenetic relationships and phylogeography of relevant lineages within the complex Campanulaceae family in Macaronesia. *Ecol. Evol.* **2017**, 1–21. [CrossRef]

76. Migliore, J.; Baumel, A.; Juin, M.; Fady, B.; Roig, A.; Duong, N.; Médail, F. Surviving in Mountain Climate Refugia: New Insights from the Genetic Diversity and Structure of the Relict Shrub *Myrtus nivellei* (Myrtaceae) in the Sahara Desert. *PLoS ONE* **2013**, *8*, e73795. [CrossRef]

77. Weigelt, P.; Steinbauer, M.J.; Cabral, J.S.; Kreft, H. Late Quaternary climate change shapes island biodiversity. *Nature* **2016**, *532*, 99. [CrossRef]

78. Romeiras, M.M.; Pena, A.R.; Menezes, T.; Vasconcelos, R.; Monteiro, F.; Paulo, O.S.; Moura, M. Shortcomings of Phylogenetic Studies on Recent Radiated Insular Groups: A Meta-Analysis Using Cabo Verde Biodiversity. *Int. J. Mol. Sci.* **2019**, *20*, 2782. [CrossRef]

79. Harter, D.E.; Irl, S.D.; Seo, B.; Steinbauer, M.J.; Gillespie, R.; Triantis, K.A.; Beierkuhnlein, C. Impacts of global climate change on the floras of oceanic islands–Projections, implications and current knowledge. *Perspect. Plant. Ecol. Evol. Syst.* **2015**, *17*, 160–183. [CrossRef]

80. Patiño, J.; Mateo, R.G.; Zanatta, F.; Marquet, A.; Aranda, S.C.; Borges, P.A.; Muñoz, J. Climate threat on the Macaronesian endemic bryophyte flora. *Sci. Rep.* **2016**, *6*, 29156. [CrossRef]

81. Monteiro, F.; Fortes, A.; Ferreira, V.; Pereira Essoh, A.; Gomes, I.; Correia, A.M.; Romeiras, M.M. Current Status and Trends in Cabo Verde Agriculture. *Agronomy* **2020**, *10*, 74. [CrossRef]

82. Caujapé-Castells, J.; García-Verdugo, C.; Marrero-Rodríguez, Á.; Fernández-Palacios, J.M.; Crawford, D.J.; Mort, M.E. Island ontogenies, syngameons, and the origins and evolution of genetic diversity in the Canarian endemic flora. *Perspect. Plant. Ecol.* **2017**, *27*, 9–22. [CrossRef]

83. Fernández-Palacios, J.M.; Arévalo, J.R.; Balguerías, E.; Barone, R.; Nascimento, L.; de Elias, R.B.; Delgado, J.D.; Fernández-Lugo, S.; Méndez, J.; Naranjo Cigala, A.; et al. *La Laurisilva. Canarias, Madeira y Azores*; Macaronesia Editorial: Santa Cruz de Tenerife, Spain, 2017.

84. Castilla-Beltrán, A.; Duarte, I.; de Nascimento, L.; Fernández-Palacios, J.M.; Romeiras, M.; Whittaker, R.J.; Jambrina-Enríquezf, M.; Mallolf, C.; Cundyg, A.B.; Edwardsa, M.; et al. Using multiple palaeoecological indicators to guide biodiversity conservation in tropical dry islands: The case of São Nicolau, Cabo Verde. *Biol. Conserv.* **2020**, *242*, 108397. [CrossRef]

85. Freitas, R.; Romeiras, M.; Silva, L.; Cordeiro, R.; Madeira, P.; González, J.A.; Wirtz, P.; Falcón, J.M.; Brito, A.; Floeter, S.R.; et al. Restructuring of the 'Macaronesia' biogeographic unit: A marine multi-taxon biogeographical approach. *Sci. Rep.* **2019**, *9*, 15792. [CrossRef] [PubMed]

 *diversity*

Review

Diversity and Conservation through Cultivation of *Hypoxis* in Africa—A Case Study of *Hypoxis hemerocallidea*

Motiki M. Mofokeng [1,2,*], Hintsa T. Araya [1,2], Stephen O. Amoo [2], David Sehlola [2], Christian P. du Plooy [2], Michael W. Bairu [2], Sonja Venter [2] and Phatu W. Mashela [1]

[1] Green Biotechnologies Research Centre of Excellence, University of Limpopo, Private Bag X1106, Sovenga 0722, South Africa; arayah@arc.agric.za (H.T.A.); phatu.mashela@ul.ac.za (P.W.M.)
[2] Agricultural Research Council, Vegetable and Ornamental Plant Private Bag X293, Pretoria 0001, South Africa; amoos@arc.agric.za (S.O.A.); dsehlola@arc.agric.za (D.S.); iduplooy@arc.agric.za (C.P.d.P.); bairum@arc.agric.za (M.W.B.); sventer@arc.agric.za (S.V.)
* Correspondence: MofokengM@arc.agric.za; Tel.: +27-(0)12-808-8000; Fax: +27-(0)12-808-0844

Received: 31 December 2019; Accepted: 20 March 2020; Published: 25 March 2020

Abstract: Africa has the largest diversity of the genus *Hypoxis*, accounting for 61% of the current globally accepted taxa within the genus, including some endemic species. Using *Hypoxis hemerocallidea* as a case study, this review addresses the conservation concerns arising from the unsustainable, wild harvesting of a number of *Hypoxis* species. *Hypoxis hemerocallidea* is one of the wild-harvested, economically important, indigenous medicinal plants of southern Africa, with potential in natural product and drug development. There are several products made from the species, including capsules, tinctures, tonics and creams that are available in the market. The use of *H. hemerocallidea* as a "cure-all" medicine puts an important harvesting pressure on the species. Unsustainable harvesting causes a continuing decline of its populations and it is therefore of high priority for conservation, including a strong case to cultivate the species. Reviewing the current knowledge and gaps on cultivation of *H. hemerocallidea*, we suggest the creation of a platform for linking all the stakeholders in the industry.

Keywords: African potato; conservation; commercialization; cultivation; Hypoxidaceae; medicinal plant; unsustainable harvesting; wild harvesting

---

1. Introduction

The African continent is known for its rich floral biodiversity, high levels of endemism [1] and increasing reliance on its natural resources based on its indigenous knowledge systems, for economic growth and development. The increasing reliance of a growing population coupled with other factors such as habitat destruction, alien species invasion and other anthropogenic factors has resulted in an increasing strain on African plant diversity, creating the need for urgent biodiversity conservation interventions [2]. Globally, there are 119 accepted taxa, including species and infra-species levels within the genus *Hypoxis* [3]. Using the latest information from the African Plant database [4], there are 72 of these taxa with accepted names [3,5] that are naturally distributed in tropical Africa, southern Africa, North Africa and Madagascar (Table 1). According to Balogun et al. [6], about 40 *Hypoxis* species are known in southern Africa, which is regarded as the main center of the species diversity and endemism [7]. Some of the species have been grouped into another genera following recent taxonomic reclassification [3,5]. About 32 *Hypoxis* species were recorded in Tropical East Africa including Uganda, Kenya, and Tanzania [8]. *Hypoxis* species are regarded as valuable medicinal plants for the treatment of numerous ailments in most parts of Africa [9]. In southern Africa, *Hypoxis hemerocallidea* is the top *Hypoxis* species with commercial value, seconded by *H. colchicifolia* [10,11]. This review therefore focuses on *H. hemerocallidea* as a case study of species within the genus although other African *Hypoxis* species of medicinal importance are also highlighted.

Hypoxis hemerocallidea Fisch., C.A.Mey. & Avé-Lall. (family: Hypoxidaceae), commonly known as African potato, iLabatheka, iNkomfe, moli and star flower, is listed in southern Africa as one of the indigenous medicinal plants with potential in natural product and drug development [12–14], hence being one of the commercially important medicinal plants in the region. It is the only *Hypoxis* species listed among the 51 plant species in the African Herbal Pharmacopoeia [15]. The commonly used part is its corms. Thus, during harvesting the corms are dug out, killing the plant [16,17]. This has most likely led to a decline in its wild populations and thus the need for conservation strategies [15]. The current review paper aims at discussing the cultivation of *H. hemerocallidea* as a conservation strategy, research work and information generated in the cultivation of the species, and recommendations going forward, which can be applicable to other *Hypoxis* species.

Hypoxis Species in Africa

The African continent alone accounts for 61% (N = 72) of the current globally accepted taxa within the genus *Hypoxis*. Some of the taxa such as *H. fischerii* var. *hockii* (De Wild.) Wiland & Nordal, *H. fischerii* var. *zernyi* (G. M. Schulze) Wiland & Nordal, *H. gregoriana* Rendle, *H. kilimanjarica* subsp. *kilimanjarica*, *H. kilimanjarica* subsp. *prostrata* E. M. Holt & Staubo and *H. urceolata* Nel are endemic to East Tropical Africa [8]. Many *Hypoxis* species are used in the management of HIV/AIDS, cancer, tuberculosis, sexually transmitted diseases and infertility (Table 1), because of their similarity in chemical constituents [18]. As a result, some of these taxa are used interchangeably in African Traditional Medicine [8]. For example, *H. angustifolia* Lam, *H. goetzei* Harms, *H. nyasica* Baker, and *H. obtusa* Burch. ex Ker Gawl. are all used in traditional medicine for similar purposes in Tanzania [7]. The corm is often the plant part used and its harvest is rather unsustainable and might lead to decimation of natural populations. At present, only *Hypoxis malaissei* Wiland has been listed in the IUCN Red List of Threatened Species [19] as 'Data Deficient'. However, this species and *H. fischeri* var. *katangensis* (De Wild.) Wiland & Nordal are suggested to be in the East Tropical Africa 'Critically Endangered' list [8]. Most species only appear on National Red List Data as 'Least Concern', but *H. fischeri* var. *colliculata* (Wiland) Wiland & Nordal, *H. kilimanjarica* Baker, *H. kilimanjarica* subsp. *Prostrata.*, and *H. polystachya* Welw. ex Baker are listed as 'Vulnerable' in the East Tropical Africa [8], and five other species are listed as 'Near Threatened' (Table 1).

Table 1. *Hypoxis* species diversity in Africa (adapted from African Plant Database [4]).

Plant Species	Synonyms of the Accepted Names	[2] Distribution in Africa	Plant Part(s) Used	Nature of Diseases/Condition Plant is Used Against	Conservation Status	Main Threats	Reference
[1] *Hypoxis abyssinica* Hochst. ex A. Rich.	*Hypoxis boranensis* Cufod.; *Hypoxis neghellensis* Cufod.; *Hypoxis petitiana* A.Rich.; *Hypoxis schweinfurthiana* Nel; *Hypoxis simensis* Hochst.; *Hypoxis tristycha* Cufod.	TA					
Hypoxis acuminata Baker		SA			LC *		
Hypoxis angustifolia Lam.	*Hypoxis angustifolia* var. angustifolia; *Hypoxis biflora* Baker	MA-SA-TA	Corm	Skin ulcers, HIV/AIDS, wounds, sickle cell diseases, ringorms	LC *		[7,8,20]
Hypoxis angustifolia var. buchananii Baker	*Hypoxis obliqua* var. woodii (Baker) Nel; *Hypoxis woodii* Baker	SA			LC *		
Hypoxis angustifolia var. luzuloides (Robyns & Tournay) Wiland	*Hypoxis luzuloides* Robyns & Tournay	MA-TA			LC @		[8]
Hypoxis angustifolia var. madagascariensis Wiland		MA					
Hypoxis argentea Harv. ex Baker	*Hypoxis argentea* var. argentea	SA-TA	Corm	Stomachache, diarrhoea	LC *		[21]
Hypoxis argentea var. sericea (Baker) Baker	*Hypoxis argentea* var. flaccida (Baker) Baker; *Hypoxis dregei* (Baker) Nel; *Hypoxis sericea* Baker; *Hypoxis sericea* var. dregei Baker; *Hypoxis sericea* var. flaccida Baker	SA-TA	Corm	Diabetes	LC *		[22]
Hypoxis kampsiana subsp. tomentosa Wiland		TA			NT @	Overgrazing, tree cutting, soil erosion, agriculture	[8]

Table 1. *Cont.*

Plant Species	Synonyms of the Accepted Names	² Distribution in Africa	Plant Part(s) Used	Nature of Diseases/Condition Plant is Used Against	Conservation Status	Main Threats	Reference
Hypoxis bampsiana Wiland	*Hypoxis bampsiana* subsp. bampsiana	TA					
Hypoxis camerooniana Baker	*Hypoxis lanceolata* Nel; *Hypoxis ledermannii* Nel; *Hypoxis petrosa* Nel; *Hypoxis recurva* (Baker) Nel; *Hypoxis thorbeckei* Nel; *Hypoxis villosa* var. recurva Baker	TA					
Hypoxis canaliculata Baker		TA					
Hypoxis colchicifolia Baker	*Hypoxis distachya* Nel; *Hypoxis gilgiana* Nel; *Hypoxis latifolia* Hook.; *Hypoxis oligotricha* Baker	SA	Corm	Diabetes, cancer, osteoporosis, purgative	LC *		[7,20–23]
		SA					
Hypoxis costata Baker		SA-TA			LC *		
Hypoxis cuanzensis Welw. ex Baker		TA					
Hypoxis decumbens L.	*Anthericum ensiforme* Vell.; *Anthericum sessile* Mill.; *Hypoxis breviscapa* Kunth; *Hypoxis caricifolia* Salisb.; *Hypoxis decumbens* var. dolichocarpa G.L.Nesom; *Hypoxis decumbens* var. major Seub.; *Hypoxis elongata* Kunth; *Hypoxis gracilis* Lehm.; *Hypoxis pusilla* Kunth; *Hypoxis racemosa* Donn.Sm.; *Niobea nemorosa* Willd. ex Schult. & Schult.f.; *Niobea pratensis* Willd. ex Schult. & Schult.f.	SA					

Table 1. *Cont.*

Plant Species	Synonyms of the Accepted Names	[2] Distribution in Africa	Plant Part(s) Used	Nature of Diseases/Condition Plant is Used Against	Conservation Status	Main Threats	Reference
Hypoxis demissa Nel		TA					
Hypoxis dinteri Nel		SA-TA					
Hypoxis exaltata Nel		SA			LC *		
Hypoxis filiformis Baker	*Hypoxis biflora* De Wild.; *Hypoxis caespitosa* Baker; *Hypoxis dregei* var. biflora Nel ex De Wild.; *Hypoxis malosana* Baker; *Hypoxis muenznerii* Nel	SA-TA			LC*, LC@		[8]
Hypoxis fischeri var. **katangensis (De Wild.) Wiland & Nordal**	*Hypoxis aculeata* Nel; *Hypoxis hockii* var. katangensis (Nel) Wiland; *Hypoxis katangensis* Nel	TA			CE @		[8]
Hypoxis fischeri var. zernyi **(G. M. Schulze) Wiland & Nordal**	*Hypoxis matengensis* G. M. Schulze; *Hypoxis zernyi* G. M. Schulze	TA	Corm	Testicular swelling	LC @		[8]
Hypoxis flanaganii Baker		SA			LC *		
Hypoxis floccosa Baker	*Hypoxis ecklonii* Baker	SA			LC *		
Hypoxis galpinii Baker	*Hypoxis infausta* Nel; *Hypoxis pungwensis* Norl.; *Hypoxis stricta* Nel	SA-TA			LC *, NT @	Near threatened in the East Tropical Africa due to disjunct distributions	[8]
Hypoxis gerrardii Baker	*Hypoxis junodii* Baker	SA	Corm	Abdominal cramps	LC *		[24]
		SA					
Hypoxis goetzei Harms	*Hypoxis esculenta* De Wild.; *Hypoxis rubiginosa* Nel; *Hypoxis turbinata* Nel	TA	Corm	Epilepsy, HIV/AIDS, stomachache	LC @		[7,8]
Hypoxis gregoriana Rendle	*Hypoxis araneosa* Nel	TA			LC @		[8]

Table 1. *Cont.*

Plant Species	Synonyms of the Accepted Names	² Distribution in Africa	Plant Part(s) Used	Nature of Diseases/Condition Plant is Used Against	Conservation Status	Main Threats	Reference
Hypoxis hemerocallidea Fisch. & Avé-Lall.	*Hypoxis elata* Hook.f.; *Hypoxis obconica* Nel; *Hypoxis patula* Nel; *Hypoxis rooperi* T.Moore; *Hypoxis rooperi* var. forbesii Baker	SA-TA	Corm	Cancer, HIV/AIDS, urinary tract diseases, reproductive system diseases, prostate hypertrophy, benign prostate hyperplasia, Tuberculosis, Syphilis, Diabetes	LC *		[8,20,25–28]
Hypoxis interjecta Nel		SA			LC *		
					LC *		
Hypoxis kilimanjarica Baker	*Hypoxis alpina* R.E.Fr.; *Hypoxis incisa* Nel; *Hypoxis kilimanjarica* subsp. kilimanjarica	TA			VU @	Human activities, does not survive in cultivated fields or after fires	[8]
Hypoxis kilimanjarica subsp. prostrata E. M. Holt & Staubo		TA			VU @	Increasing tourist activities	[8]
Hypoxis kraussiana Buchinger ex C. Krauss		SA			LC *		
Hypoxis lata Nel		SA			LC *		
Hypoxis lejolyana Wiland		TA					
Hypoxis leucotricha Fritsch		TA					
Hypoxis limicola B. L. Burtt		SA			LC *		
Hypoxis longifolia Baker	*Hypoxis longifolia* var. thunbergii Baker	SA	Corm	Gynaecology and obstetric disorders	LC *		[29]

Table 1. *Cont.*

Plant Species	Synonyms of the Accepted Names	[2] Distribution in Africa	Plant Part(s) Used	Nature of Diseases/Condition Plant is Used Against	Conservation Status	Main Threats	Reference
Hypoxis ludwigii Baker		SA			LC *		
Hypoxis lusalensis Wiland		TA					
Hypoxis malaissei Wiland		TA			Data Deficient [#], CE [@]		[8]
Hypoxis membranacea Baker		SA			LC *		
Hypoxis monanthos Baker		TA					
Hypoxis muhilensis Wiland	*Hypoxis muhilensis* subsp. muhilensis	TA					
Hypoxis multiceps Buchinger ex Baker		SA-TA			LC *		
Hypoxis neliana Schinz		SA			LC *		
Hypoxis nyasica Baker	*Hypoxis campanulata* Nel; *Hypoxis engleriana* Nel; *Hypoxis engleriana* var. scottii Nel; *Hypoxis ingrata* Nel; *Hypoxis probata* Nel; *Hypoxis retracta* Nel	TA	Corm	Coughs, HIV/AIDS	LC [@]		[8]
Hypoxis oblonga Nel		SA			LC *		
Hypoxis obtusa Burch. ex Ker Gawl.	*Hypoxis angolensis* Baker; *Hypoxis iridifolia* Baker; *Hypoxis nitida* Verd.; *Hypoxis obtusa* var. chrysotricha Nel; *Hypoxis villosa* var. obtusa (Burch. ex Ker Gawl.) T.Durand & Schinz	SA-TA	Corm	Diabetes, ulcers, HIV/AIDS, abdominal pains	LC *, NT [@]	Near threatened in the East Tropical Africa due to disjoint distribution, and regular gathering for herbarium collections	[7,8,30]

Table 1. *Cont.*

Plant Species	Synonyms of the Accepted Names	[2] Distribution in Africa	Plant Part(s) Used	Nature of Diseases/Condition Plant is Used Against	Conservation Status	Main Threats	Reference
Hypoxis parvifolia **Baker**		SA-TA			LC *		
Hypoxis parvula **Baker**	*Hypoxis brevifolia* Baker; *Hypoxis parvula* var. parvula	SA			LC *		
Hypoxis polystachya **Welw. ex Baker**	*Hypoxis orbiculata* Nel; *Hypoxis polystachya* var. andongensis Baker; *Hypoxis subspicata* Pax	TA	Corm	Testicular swelling, fungal infection of scalp	VU@	Strong antropogenic pressure, collection for medicinal purpose	[8]
Hypoxis protrusa **Nel**		TA					
Hypoxis rigidula **Baker**	*Hypoxis cordata* Nel; *Hypoxis elliptica* Nel; *Hypoxis laikipiensis* Rendle; *Hypoxis rigidula* var. rigidula; *Hypoxis volkmanniae* Dinter	SA-TA	Corm	Gall sickness	LC *, LC @		[8]
Hypoxis rigidula var. **pilosissima Baker**	*Hypoxis arnottii* Baker	SA-TA			LC *		
Hypoxis robusta **Nel ex De Wild.**		TA					
Hypoxis sagittata **Nel**		SA			LC *		
Hypoxis schimperi **Baker**	*Hypoxis macrocarpa* E.M.Holt & Staubo	TA	Corm	Coughs	NT @	Limited and scattered island-like distribution; possibility of being used as a substitute for other more commonly used *Hypoxis* species	[7,8]

Table 1. *Cont.*

Plant Species	Synonyms of the Accepted Names	[2] Distribution in Africa	Plant Part(s) Used	Nature of Diseases/Condition Plant is Used Against	Conservation Status	Main Threats	Reference
Hypoxis setosa Baker		SA			LC *		
Hypoxis sobolifera Jacq.	*Hypoxis canescens* Fisch. & C.A.Mey.; *Hypoxis krebsii* Fisch.; *Hypoxis pannosa* Baker; *Hypoxis sobolifera* var. accedens Nel; *Hypoxis sobolifera* var. pannosa (Baker) Nel; *Hypoxis villosa* var. canescens (Fisch. & C.A.Mey.) Baker; *Hypoxis villosa* var. pannosa (Baker) Baker; *Hypoxis villosa* var. sobolifera (Jacq.) Baker	SA	Corm	Immune boosting	LC *		[31]
Hypoxis stellipilis Ker Gawl.	*Hypoxis lanata* Eckl. ex Baker	SA	Corm	Immune boosting	LC *		[31]
Hypoxis suffruticosa Nel		TA					
Hypoxis symoensiana Wiland		TA					
Hypoxis tetramera Hilliard & E. L. Burtt		SA			LC *		
Hypoxis uniflorata Markötter		SA			DDT *		
Hypoxis upembensis Wiland		TA					
Hypoxis urceolata Nel	*Hypoxis apiculata* Nel; *Hypoxis arenosa* Nel; *Hypoxis bequaertii* De Wild.; *Hypoxis crispa* Nel; *Hypoxis cryptophylla* Nel; *Hypoxis textilis* Nel	TA			LC @		[8]

Table 1. *Cont.*

Plant Species	Synonyms of the Accepted Names	[2] Distribution in Africa	Plant Part(s) Used	Nature of Diseases/Condition Plant is Used Against	Conservation Status	Main Threats	Reference
Hypoxis villosa L. f.	*Fabricia villosa* Thunb.; *Hypoxis decumbens* Lam. [Illegitimate]; *Hypoxis fabricia* Gaertn.; *Hypoxis microsperma* Avé-Lall.; *Hypoxis obliqua* Jacq.; *Hypoxis scabra* Lodd.; *Hypoxis tomentosa* Lam.; *Hypoxis villosa* var. fimbriata Nel; *Hypoxis villosa* var. obliqua (Jacq.) Baker; *Hypoxis villosa* var. scabra (Lodd.) Baker; *Hypoxis villosa* var. villosa	SA-TA			LC *		
Hypoxis zeyheri Baker		SA			LC *		

[1] Boldly written taxa are accepted names according to The Plant List [5] and World Checklist of Selected Plant Families [3]. [2] Distribution in Africa: TA: Tropical Africa Area (EPFAT Area, country-based, south of the Sahara, complementary to the following); SA: Southern Africa Area (South Africa, Namibia, Botswana, Lesotho, Swaziland); NA: North Africa (Mauritania, Morocco, Canary Isl., Algeria, Tunisia, Libya, Egypt, Madeira); MA: Madagascar (Malagasy Republic). [#] Global conservation status according to IUCN [32]. * Conservation status according to The Red List of South African Plants (SANBI) [33]; LC: Least Concern; DDT: Data Deficient—Taxonomically Problematic; VU: Vulnerable. [@] East Tropical Africa Conservation status [8]; CE: Critically Endangered; VU: Vulnerable; NT: Near Threatened; LC: Least Concern.

2. Natural History and Distribution of *Hypoxis hemerocallidea*

Hypoxis hemerocallidea, previously known as *H. rooperi* [3,5], is a perennial corm with long, broad, and slightly hairy leaves [34,35] (Figure 1). Some species have a close morphological resemblance with *H. hemerocallidea*, such as *H. acuminata* Baker, *H. colchifolia* Baker, *H. galpinii* Baker, and *H. obtusa* [36]. The long, hairy and sickle shaped leaves of *H. hemerocallidea* differentiate it from the other species [36].

(a)   (b)   (c)

Figure 1. The yellow flowers (**a**), hairy leaves and black seeds (**b**) of *Hypoxis hemerocallidea* and its corm (**c**).

Hypoxis hemerocallidea has yellow star-shaped flowers [36], of which between one to three open within an interval of an hour per day, a strategy to encourage cross-pollination [37]. Although *H. hemerocallidea* plants produce many seeds, very low germination percentages were reported [38], due to the hard, shiny, black seed coat which may be impermeable to water and may restrict oxygen movement to the embryo, thus delaying germination [37,38]. Most mature corms, which are dark-brown to black outside and yellow inside, are approximately 10–15 cm in diameter and about half a kilogram in weight [39]. The fleshy corm is a survival mechanism that enables the plant to withstand cold conditions, drought, and veld fire [40]. In South Africa, *H. hemerocallidea* grows naturally in the savanna grasslands in KwaZulu-Natal, Eastern Cape, Mpumalanga, Limpopo, Gauteng, and Free State Provinces, and also grows in other countries such as Lesotho, Swaziland, Mozambique, and Zimbabwe [41].

3. Uses and Chemical Composition of *Hypoxis hemerocallidea*

Hypoxis hemerocallidea is traditionally used as a tonic, purgative, diuretic or to treat infertility, inflammation, prostate gland disorder, wounds and burns [14,42]. Promising anticancer activities of *H. hemerocallidea* have also been reported [43], which support the listing of *H. hemerocallidea* as one of the species used for the treatment of cancer in the Eastern Cape Province of South Africa [44]. Furthermore, it is also used for the treatment of diabetes in Eastern Cape and Eastern Free State [6,16]. The species was made famous by the South African Health Ministry when it was recommended as an immune booster for people living with HIV/AIDS [16,44,45]. The 14 member states of the Southern African Development Community (SADC) supported the use of 'African potato' in HIV management [46]. *Hypoxis hemerocallidea* was found to be a good source of trace elements such as zinc, copper, and manganese, which could account for its use as a pro-fertility ingredient and for boosting the immune system [47]. However, there are warnings of a potential herb-drug-interaction if *H. hemerocallidea* is used with conventional drugs for the management of HIV/AIDS [45]. It was also mentioned as one of the three most used plant species in the treatment of sexually transmitted diseases, in KwaZulu-Natal and Limpopo Provinces of South Africa [48,49]. In support of its traditional use as an anti-miscarriage medicine, apart from being used for bronchial asthma [50], a study by Nyinawumuntu et al. [51] showed that *H. hemerocallidea* corm extracts possess uterolytic activity. *Hypoxis hemerocallidea* is mentioned as one of the species used for the treatment of tuberculosis in Limpopo Province of South Africa [52]. The species was one of the most cited medicinal plants in the medicinal markets [53], used for treatment of malaria and venereal diseases in Mozambique [26,54]. Different *Hypoxis* spp. are used in the treatment of prostate cancer in Zimbabwe [55]. A description of the history, first research reports and prominent promotion of commercial uses of *H. hemerocallidea* [43] is adapted in Figure 2. Traditional uses and chemical constituents of 11 *Hypoxis* species, including 14 *H. hemerocallidea* herbal formulations indicating the commercial importance of this species, are summarized in other reports [56].

A glycoside called hypoxoside is mentioned as the main component of *H. hemerocallidea*, with sterols (stigmasterol, β-sitosterol, campesterol), sterolins, norlignan, daucosterols and stanols (stigmastanol) as additional constituents [42,46,57–59]. Amongst these phytochemicals, daucosterols, β-sitosterol, and hypoxide are associated with the plants' therapeutic activities [60]. Phenylalanine and *t*-cinnamic acid are reported to be precursors of hypoxoside in whole *H. hemerocallidea* plants [61]. Although some studies have proved that *Hypoxis* extracts with 45% hypoxoside are not toxic [41], even after long-term use of *H. hemerocallidea* products [56], treatment of diabetic rats with a high dose of 800 mg/kg resulted in abnormal kidney function [58].

Hypoxis hemerocallidea, H. rigidula Baker, *H. galpinii,* and *H. obtusa* showed similar phytochemical profiles, although further studies were recommended to confirm their safety [60]. A higher concentration of ergosterol and stigmasterol was reported in *H. rigidula* when compared to *H. hemerocallidea* [62]. When *H. hemerocallidea* was compared to *H. stellipilis* Ker Gawl. and *H. sobolifera* Jacq., it was found that both sterol and hypoxoside contents varied amongst the species, which questions the indiscriminate use of *Hypoxis* spp. in traditional medicine [60]. Several studies have been conducted to investigate the hypoglycaemic effects [63,64], typhlocolitic effects [65] and its effects on pharmacokinetics of efavirenz [57], antibiotic and immune modulation phytotherapies [66].

The common name 'African potato' is misleading, as many perceive *H. hemerocallidea* as edible. *Hypoxis hemerocallidea* has been reported as one of the indigenous and traditional food crops consumed in the North West Province of South Africa [67]. This species could be confused with the edible *Plectranthus esculentus* N.E.Br. as they both share the same common name of 'African potato' [67]. *Hypoxis hemerocallidea* has low crude protein value and low accumulation of selected elements [68], compared to human daily requirement, but its high iron content makes it a good candidate for use in overcoming iron deficiencies [47]. Some geophytes of *Hypoxis* were found in the caves in South Africa, evidence for cooking of edible rhizomes centuries ago [69].

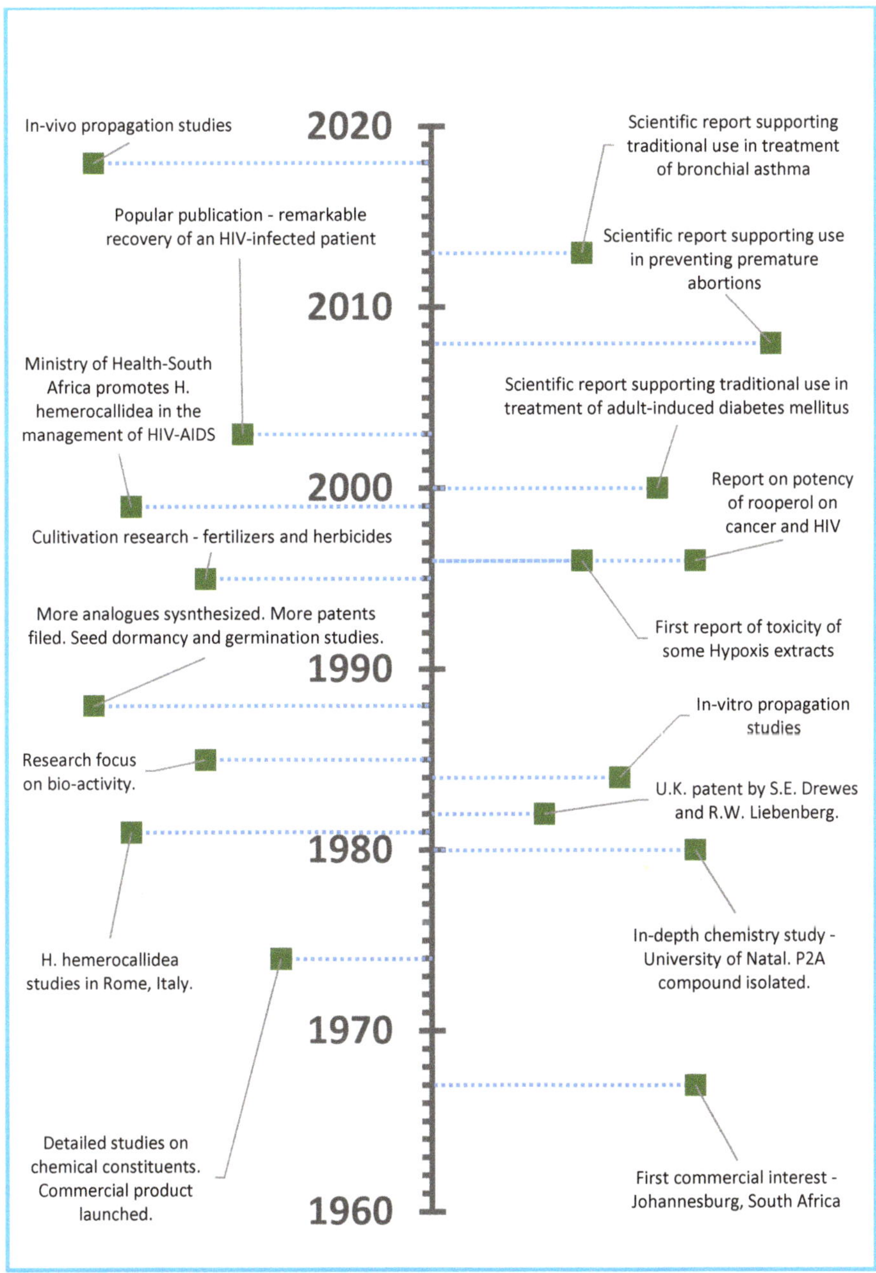

Figure 2. The timeline (1960–2020) of research and commercial use of *Hypoxis hemerocallidea*.

4. Market Demand of *Hypoxis hemerocallidea*

Products incorporating extracts of medicinal plants such as *H. hemerocallidea*, which were used solely in traditional medicine are increasinlgy becoming commercialized [70]. The first *Hypoxis hemerocallidea* product was developed in 1967, in Johannesburg, South Africa and was successfully marketed in Germany [71]. An increase in the market share of *H. hemerocallidea* and *H. colchifolia* in 2001 was attributed to the publicity of the species received after several media releases proclaiming its healing properties [72]. Even before this increase in consumption, it was found that 77% of the *Hypoxis* corms sold in the Witwatersrand informal medicinal plant markets were 2–7 cm in diameter, with traders compensating for the smaller size by selling a large number of smaller corms [72]. This could be an indication of the depletion of bigger corms in the wild, leading to smaller immature bulbs being harvested. For example, the size of *Eucomis autumnalis* (Mill.) Chitt bulbs sold in the market, decreased in size between 1995 and 2001, indicating the negative impacts of harvesting for trade [72]. Nevertheless, pressure on this resource continues as *Hypoxis* spp. (*H. hemerocallidea* and *H. colchifolia*) were ranked fourth in terms of volumes sold in 2007, at the Faraday market in Johannesburg [72] and was one of the 19 species that had a high demand in Zululand market [73]. 'African potato' was one of the top 10 most frequently sold medicinal plant species, reaching 11,000 kg/annum to the value of R 322,500 (approximately 21,930 USD; at the 2019 average rate of R 1.00 = 0.068 USD), in Eastern Cape Province of South Africa [74]. Approximately 31,300 corms of *H. hemerocallidea* were sold annually from 54 outlets in Durban, South Africa [75]. A factory in KwaZulu-Natal (South Africa) was reported in 2004 to be processing 1.5 tons of *H. hemerocallidea* derived products and this was expected to increase annually [29]. In Lesotho, *H. hemerocallidea* was mentioned as one of the medicinal plants frequently used for a wide range of ailments [76]. This plant was the second most commonly used species in Kimberly, South Africa [77]. In Maputo, Mozambique, *H. hemerocallidea* topped the list of medicinal plant species mentioned by 71% of traders interviewed [78]. Figure 3 shows *H. hemerocallidea* corms displayed at an informal market in northern KwaZulu-Natal, South Africa.

Many products made from *H. hemerocallidea*, which include capsules, tinctures, tonics and creams, are available in the market [79]. *Hypoxis hemerocallidea* is exported, in a preprocessed form, to Asia and Europe, where the extracts are used for treatment of prostate problems [14]. In a review on medicinal plants with potential in the development of drugs, *H. hemerocallidea* is mentioned as one of the famous African medicinal plants [80] with an ever-increasing demand.

Figure 3. *Hypoxis hemerocallidea* corms displayed (red arrow at middle bottom of the picture) at an informal market in northern KwaZulu-Natal, South Africa, with other medicinal plants.

5. Conservation Status of *Hypoxis hemerocallidea*

The conservation status of *H. hemerocallidea* is described as declining, and although not endangered with extinction, harvesting would cause a continuing decline of its populations and it is therefore of priority for conservation [40,81], with frequent monitoring warranted [74]. The only conservation effort was reported in KwaZulu-Natal, South Africa, where 250 hectares of land has been reserved for protection of *H. hemerocallidea* [81].

The use of *H. hemerocallidea* as an "all-purpose", "cure-all", or "wonder plant" medicine [37,50] has put pressure on wild populations because it presents an opportunity for income generation, through species-specific trade network [56]. During harvesting, the whole plant is uprooted, resulting in total destruction of the plant [75]. There is a need for urgent propagation intervention [73] as their populations in the wild are declining [1]. For example, in Swaziland, *H. hemerocallidea* has become threatened with extinction [82]. In an effort to come up with sustainable harvesting strategies, Katerere and Eloff [83] found that *H. hemerocallidea* leaves cannot be used as a substitute for the corms due to lack of similarities in phytochemical content and bioactivity. However, it was suggested that the leaves of *H. hemerocallidea* might be used as substitutes for the corm in the treatment of some bacterial and fungal diseases as there seemed to be a shift in activity between plant organs in different seasons [56]. Hypoxoside was reported to be hydrolysed in the leaves and then transported to the corms [61], with this activity shifting from the leaves to the corms in summer [56]. Du Toit et al. [84] reported that harvesting season affected the concentration of active ingredients produced from the corms during the year. Therefore, the correct timing of collection of *H. hemerocallidea* leaves could ease the pressure off wild populations. The activity was higher in the leaves in spring [56], a period in which the corms are regrowing the leaves after a period of dormancy when there are fewer or smaller leaves available. In summer, when there is a greater number of bigger leaves, the activity is shifted to the corms and this could be another reason for continued use of the corms than the leaves.

6. Propagation and Cultivation of *Hypoxis hemerocallidea*

Due to the value of *H. hemerocallidea* in traditional medicine and the likely diminishing wild populations, there is a strong case to cultivate the species for sustainable supply of good quality corms [35,45,75], agreeing with Katerere and Eloff [83] who mentioned that the only viable alternative to unsustainable wild harvesting of *H. hemerocallidea* is propagation and domestication. In essence, cultivation will not only alleviate pressure on natural resources, but it also has practical implications on optimising secondary metabolite production, facilitating standardisation, and increased safety by reducing inconsistency in quality and composition of plant material, reducing the risk of adulteration and increasing yield through management practices [85]. Encouragement of cultivation of wild harvested medicinal plants and development of nurseries to propagate the species will take off the pressure on wild stocks [44]. Cultivation of medicinal plants is becoming popular, with increasing government support to meet the high demands for plant materials [86]. The bioprospecting economy and related strategies in South Africa have put an important drive to mass propagation and cultivation of commercially important medicinal plants, of which *H. hemerocallidea* is within the top 25. However, one of the major factors hindering the improvement in cultivation of *H. hemerocallidea* is lack of knowledge of its agronomic and quality traits [40].

Propagation of *H. hemerocallidea* was identified as problematic since seed dormancy is difficult to be broken and the species does not propagate easily from corms; thus studies relating to reproductive biology and seed physiology of the species were recommended as a priority [12,40]. Seed germination of *H. hemerocallidea* is also a barrier because of unpredictable seed viability, combinational dormancy, poor seedling establishment, while vegetative propagation is rare [38]. Micropropagation of *H. hemerocallidea* is the only vegetative propagation method that has been thoroughly studied by many researchers [87]. Corm explants of *H. hemerocallidea* seem to take at least nine weeks to respond positively in in vitro culture compared to other *Hypoxis* species [88]. Other challenges with in vitro propagation of *H. hemerocallidea* include browning of explants and nutrient medium which necessitates regular subculturing of the shoots at six weeks' interval [81]. Endogenous contamination was also reported as a major challenge when underground organs were used as explants [75]. Ndong et al. [40] explored the regeneration potential of *H. hemerocallidea* corm explants by inoculating them on culture media with a wide range of plant growth regulator combinations and concentrations. The study concluded that no shoots were formed without plant growth regulators; shoot formation frequency and intensity were mainly dependent on cytokinin concentration and the best results were obtained with kinetin than with benzylaminopurine (BA). In a similar study, Nsibande et al. [89], reported that corm explants of *H. hemerocallidea* were very slow and relatively poor in responding to in vitro culture, with the majority of the treatments forming no callus, compared to other three *Hypoxis* spp. (*H. filiformis*, *H. argentea*, and *H. acuminata*). Direct shoot development started 10 weeks after transferring of the corm explants to a Murashige and Skoog medium supplemented with cytokinins, although growth of the shoots was prolonged [89]. Cytokinins have been reported to have a promoting effect on shoot formation in tuberous plants including *Hypoxis* spp. [40]. The stimulation of shoot regeneration, proliferation and growth by cytokinins is a species-dependent phenomenon, which is influenced by structure-activity relationship, cytokinin homeostasis, concentration of inherent endogenous plant growth regulators and auxin-cytokinin cross-talk [81]. Somatic embryogenesis, which involves the development of embryos from normal plant tissue, with the use of 2,4-D and BA produced a high number of somatic embryos while GA$_3$ was successful in developing the somatic embryos into plantlets [79].

Seedling emergence of *H. hemerocallidea* was observed a year after sowing [38]. Gillmer and Symmonds [90] have successfully germinated a large number of *H. hemerocallidea* seed by watering them three times in a day. In other experiments, there was no germination of intact seeds of *H. hemerocallidea* at three different temperatures (20, 25, and 30 °C), even when the seeds were treated with several chemicals [38]. Complete removal of the seed coat with incubation at 20 °C under white light resulted in 36% germination compared to dark conditions where only 6% germinated [38]. Seed coat seem to inhibit prompt germination of *H. hemerocallidea* seeds. Gibberellic acid (GA$_3$) at 10^{-3} mol dm^{-3} increased germination of coatless seeds, at 25 °C [37], as well as mechanical scarification and subsequent soaking for 24 h in 50 mL of GA$_3$ at 1200 ppm, which increased germination to 60% [38]. The scarification could have removed the physical barrier, while the GA$_3$ could have broken the embryo dormancy. The positive effect of GA$_3$ further suggests the presence of embryo dormancy in addition to the physical dormancy. For example, seeds harvested in early November (early summer) from Mountain Rise in KwaZulu-Natal (KZN), South Africa, reached a germination percentage of 52% at 25 °C, compared to 16% and 9% achieved from seeds harvested at the same site in early December and early January (mid-summer) [35]. The study by Hammerton et al. [35] suggests that microclimates and differences in soil conditions from different sites may have affected seed viability and dormancy. Seeds collected from a second site (Hayfields in KZN) in late December only achieved approximately 13% germination [35].

Although work has been done on in vitro propagation and seed germination of *H. hemerocallidea*, there is a need to develop methods that can supply the market with the material at reasonable costs [41]. In an effort to develop an easy and affordable propagation method for *H. hemerocallidea*, chipping into equal segments and scooping to remove the growth tip were found to be effective propagation methods [87].

Care is also needed as *H. hemerocallidea* plants to which herbicides were not applied grew better and had a higher hypoxoside content, compared to herbicide-treated plants [91]. Competition with weeds could have led to plant stress and thus an increase in secondary metabolite synthesis. It seems that high levels of nutrients (N, P, and K) are only required initially to produce a suitable *H. hemerocallidea* biomass and after plant establishment, fertilizer treatment can be discontinued [91]. Although *H. hemerocallidea* prefers poor soils with little nutrients, site selection for cultivation of the species is essential as exposure to cadmium (Cd) and aluminium (Al) significantly decreased the accumulation of hypoxoside [86]. Du Toit et al. [84] found that growing media and harvesting season affected the amount of compounds or active ingredients produced from the corms during the year.

7. Gaps in Cultivation Research of *Hypoxis hemerocallidea*

Currently, cultivation information, including the response of *H. hemerocallidea* to agronomic practices, is limited to few studies. Although *H. hemerocallidea* seems to be a drought-tolerant species, there is no official record of it growing in the Northern Cape and Western Cape Provinces of South Africa (Figure 4). According to the South African National Biodiversity Institute (SANBI), the species mostly grows in areas having 600 to 1000+ mm/annum rainfall, which includes the semi-arid and dry-sub humid zones of South Africa, with warm temperate climate, cool to hot summers, and humid to dry winters [92]. This indicates a need to investigate watering requirements from the seedling establishment stage to maturity. Imposing water stress at specific growth stages or even for a specific period before harvesting could increase growth while ensuring good content of active compounds. Ideally, cultivation at a commercial scale, provides an opportunity for manipulation of growing conditions to shorten the period to reach maturity, increase potency and predictability of extracts. *Pelargonium sidoides* DC. plants grown under greenhouse conditions, for example, showed equivalent concentrations of the compound of interest, umckalin, and approximately six times greater growth rates [93]. Excessive watering (twice and thrice a week) of *Dioscorea dregeana* (Kunth) T. Durand & Schinz seedlings reduced vegetative growth but significantly improved seedling weight and tuber size, while watering once a week resulted in better above ground biomass [94].

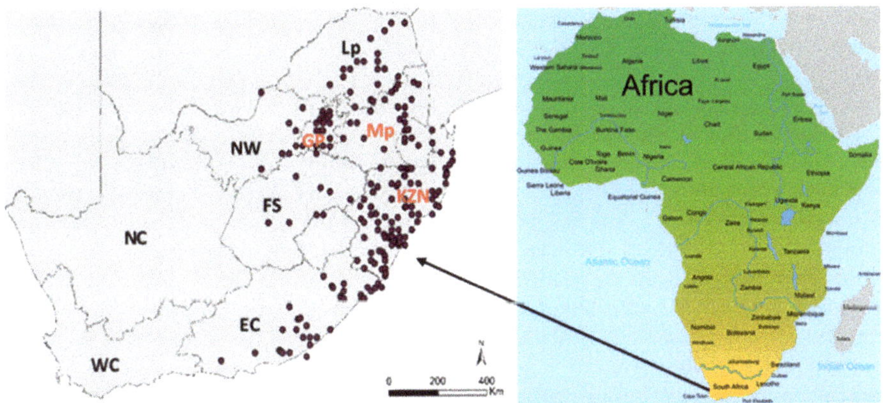

Figure 4. Distribution of *Hypoxis hemerocallidea* in South Africa, showing no records in Northern (NC) and Western Cape (WC) Provinces (Source: South African National Biodiversity Institute; SANBI). Provinces: WC: Western Cape; NC. Northern Cape; EC: Eastern Cape; FS: Free State; NW: North West; GP: Gauteng; KZN: KwaZulu-Natal; Mp: Mpumalanga; Lp: Limpopo.

The effect of fertilizers or exposure to different nutrient status on growth and medicinal activity of *H. hemerocallidea* is not fully understood. For example, in *Siphonochilus aethiopicus* (Schweif.) B.L. Burt, plants exposed to low nitrogen levels with severe water stress showed increased flavonoids, phenolics, and antioxidants indicating that fertilizer application and irrigation can alter secondary metabolite content [95]. Understanding of the strategies employed by *H. hemerocallidea* under water and nutrient stress conditions in relation to its growth, physiology, and secondary metabolite production is important.

Furthermore, field observations by Mofokeng et al. [96] showed that cultivated *Pelargonium sidoides* was susceptible to root-knot nematodes, such that root yield was negatively affected. The quality of *H. hemerocallidea* corms could be negatively affected if the species can be found to be susceptible to root-knot nematodes. Figure 5 shows a potential pest (African bollworm: *Helicoverpa armigera* Hübner) of *H. hemerocallidea*, feeding on the leaves and seeds, observed in a small scale plantation.

Other important aspects of commercial cultivation of *H. hemerocallidea* include optimum size and timing before harvesting the species, as it is not clear at what age or size are the medicinally active compounds available or optimum in the corm. In general, the target compounds in medicinal plants are secondary metabolites which serve as an adaptation strategy to environmental stress, infection, or herbivory. Thus, agronomic practices that can manipulate the growth and synthesis of secondary metabolites in *H. hemerocallidea* need to be fully understood if the species is to be successfully cultivated.

Figure 5. A potential pest and its potential damage on *Hypoxis hemerocallidea* plants. (**A**) African bollworm (*Helicoverpa armigera*) feeding on the leaves and (**B**) on the seed pods (Source: Agricultural Research Council-Vegetable and Ornamental Plant Research Station).

8. Conclusions

The decline in wild populations and the increasing demand for *H. hemerocallidea* present an opportunity to cultivate the species for conservation purposes but also as an alternative cash crop for different communities. Sustainable harvesting methods and the use of leaves as an alternative to corms should be further researched, including increasing leaf biomass and manipulating active compound content in the leaves. The main challenges for cultivation of *H. hemerocallidea*, as with many perennial medicinal plants, are the relatively long period before harvesting, in competition with wild harvest, which may reduce prices, and spatial-temporal competition with food crops. Opportunities are increasing since government legislation on wild harvesting is becoming stricter; as the demand for natural health products is also increasing, and with cultivation, pharmaceutical companies are assured of reliable supply of quality material. For example, the South African government has started to promote cultivation of commercially important indigenous medicinal plants and created a platform for linking all stakeholders in the industry. Furthermore, funding for initial investment, which is a big problem for rural poor farmers, can be made available through the abovementioned platform. As the current wild harvesting of *H. hemerocallidea* is probably unsustainable and considering its substitution with other *Hypoxis* species, it is imperative that population studies of many African *Hypoxis* species and their conservation status be reviewed for more information in order to facilitate timeous conservation and management strategies.

Author Contributions: Conceptualization, M.M.M. and H.T.A.; writing—original draft preparation, M.M.M., D.S., and H.T.A.; writing—review and editing, S.O.A., S.V., and M.W.B.; supervision, C.P.d.P. and P.W.M. All authors have read and agreed to the published version of the manuscript.

Funding: This research received no external funding.

Conflicts of Interest: The authors declare no conflict of interest.

References

1. Moyo, M.; Aremu, A.O.; Van Staden, J. Medicinal plants: An invaluable, dwindling resource in sub-Saharan Africa. *J. Ethnopharmacol.* **2015**, *174*, 595–606. [CrossRef] [PubMed]
2. Van Wyk, A.S.; Prinsloo, G. Medicinal plant harvesting, sustainability and cultivation in South Africa. *Biol. Conserv.* **2018**, *227*, 335–342. [CrossRef]
3. World Checklist of Selected Plant Families. Available online: https://wcsp.science.kew.org/namedetail.do?name_id=278949 (accessed on 30 January 2020).

4. African Plant Database. (Version 3.4.0). Conservatoire et Jardin botaniques de la Ville de Genève and South African National Biodiversity Institute, Pretoria, 2012. Available online: http://www.ville-ge.ch/musinfo/bd/cjb/africa/ (accessed on 25 December 2019).

5. The Plant List. Version 1.1. Available online: http://www.theplantlist.org/ (accessed on 25 December 2019).

6. Balogun, F.O.; Tshabalala, N.T.; Ashafa, A.O.T. Antidiabetic medicinal plants used by Basotho tribe of Eastern Free State: A review. *J. Diabetes Res.* **2016**, 1–13. [CrossRef]

7. Pereus, D.; Otieno, J.N.; Ghorbani, A.; Kocyan, A.; Hilonga, S.; de Boer, H.J. Diversity of *Hypoxis* species used in ethnomedicine in Tanzania. *S. Afr. J. Bot.* **2019**, *122*, 336–341. [CrossRef]

8. Wiland-Szymańska, J. The genus *Hypoxis* L. (Hypoxidaceae) in the East Tropical Africa: Variability, distribution and conservation status. *Biodivers. Res. Conserv.* **2009**, *14*, 1–129. [CrossRef]

9. Nsibande, B.E.; Gustavsson, K.-E.; Zhu, L.-H. Analysis of health-associated phytochemical compounds in seven *Hypoxis* species. *Am. J. Plant Sci.* **2018**, *9*, 571–583. [CrossRef]

10. Appleton, M.R.; Ascough, G.D.; Van Staden, J. In vitro regeneration of *Hypoxis colchicifolia* plantlets. *S. Afr. J. Bot.* **2012**, *80*, 25–35. [CrossRef]

11. Cumbe, J.T. Antidiabetic Compounds from *Hypoxis colchicifolia* and *Terminalia sericea*. Ph.D. Thesis, University of KwaZulu-Natal, Pietermaritzburg, South Africa, 2015.

12. Van Wyk, B.-E. A broad review of commercially important southern African medicinal plants. *J. Ethnopharmacol.* **2008**, *119*, 342–355. [CrossRef]

13. Williams, V.L.; Balkwill, K.; Witkowski, E.T.F. Size-class prevalence of bulbous and perennial herbs sold in the Johannesburg medicinal plant markets between 1995 and 2001. *S. Afr. J. Bot.* **2007**, *73*, 144–155. [CrossRef]

14. Hostettmann, K.; Marston, A.; Ndjoko, K.; Wolfender, J.L. The potential of African plants as a source of drugs. *Curr. Org. Chem.* **2000**, *4*, 973–1010. [CrossRef]

15. Sathekge, N.R. Comparison of Secondary Metabolite Content and Antimicrobial Activity of four *Hypoxis* Species Used in Traditional Medicine. Ph.D. Thesis, University of Pretoria, Pretoria, South Africa, 2010.

16. Erasto, P.; Adebola, P.O.; Grierson, D.S.; Afolayan, A.J. An ethnobotanical study of plants used for the treatment of diabetes in the Eastern Cape Province, South Africa. *Afr. J. Biotechnol.* **2005**, *4*, 1458–1460.

17. Malungu, N. Self-reported use of traditional, complementary and over-the-counter medicines by HIV-infected patients on antiretroviral therapy in Pretoria, South Africa. *Afr. J. Tradit. Complement. Altern. Med.* **2007**, *4*, 273–278. [CrossRef] [PubMed]

18. Bassey, K.; Viljoen, A.; Combrinck, S.; Choi, Y.H. New phytochemicals from the corms of medicinally important South African Hypoxis species. *Phytochem. Lett.* **2014**, *10*, lxix–lxxv. [CrossRef]

19. Eastern Arc Mountains & Coastal Forests CEPF Plant Assessment Project 2009. *Hypoxis malaissei*. The IUCN Red List of Threatened Species 2009, e.T158059A5181441. Available online: http://dx.doi.org/10.2305/IUCN.UK.2009-2.RLTS.T158059A5181441.en (accessed on 27 December 2019).

20. Mpiana, P.T.; Ngbolua, K.N.; Mudogo, V.; Tshibangu, D.S.T.; Atibu, E.K.; Mbala, B.M.; Kahumba, B.; Bokota, M.T.; Makelele, L.T. The potential effectiveness of medicinal plants used for the treatment of sickle cell disease in the Democratic Republic of Congo Folk medicina: A review. *Progress Tradit. Folk Herb. Med.* **2012**, *1*, 1–11.

21. Wintola, O.A.; Otang, W.M.; Afolayan, A.J. The prevalence and perceived efficacy of medicinal plants used for stomach ailments in the Amathole district municipality, Eastern Cape, South Africa. *S. Afr. J. Bot.* **2017**, *108*, 144–148. [CrossRef]

22. Odeyemi, S.; Bradley, G. Medicinal plants used for the traditional management of diabetes in the Eastern Cape, South Africa: Pharmacology and toxicology. *Molecules* **2018**, *23*, 2759. [CrossRef]

23. Madikizela, B.; McGaw, L.J. In vitro cytotoxicity, antioxidant and anti-inflammatory activities of *Pittosporum viridiflorum* and *Hypoxis colchicifolia* Baker used traditionally against cancer in Eastern Cape, South Africa. *S. Afr. J. Bot.* **2019**, *126*, 250–255. [CrossRef]

24. Aremu, A.O.; Ndhlala, A.R.; Fawole, O.A.; Light, M.E.; Finnie, J.F.; Van Staden, J. In vitro pharmacological evaluation and phenolic content of ten South African medicinal plants used as anthelmintics. *S. Afr. J. Bot* **2010**, *76*, 550–566. [CrossRef]

25. Amusan, O.O.G.; Sukati, N.A.; Dlamini, P.S.; Sibandze, F.G. Some Swazi phytomedicines and their constituents. *Afr. J. Biotechnol.* **2007**, *6*, 267–272.

26. Bandeira, S.O.; Gaspar, F.; Pagula, F.P. African ethnobotany and healthcare: Emphasis on Mozambique. *Pharm. Biol.* **2001**, *1*, 70–73.

27. Bhat, R.B. Medicinal plants and traditional practices of Xhosa people in the Transkei region of Eastern Cape, South Africa. *Indian J. Tradit. Know.* **2014**, *13*, 292–298.

28. Kose, L.S.; Moteetee, A.; Van Vuuren, S. Ethnobotanical survey of medicinal plants used in the Maseru district of Lesotho. *J. Ethnopharmacol.* **2015**, *170*, 184–200. [CrossRef] [PubMed]

29. Drewes, S.E.; Elliot, E.; Khan, F.; Dhlamini, J.T.B.; Gcumisa, M.S.S. *Hypoxis hemerocallidea*–not merely a cure for benign prostate hyperplasia. *J. Ethnopharmacol.* **2008**, *119*, 593–598. [CrossRef] [PubMed]

30. De Wet, H.; Ngubane, S.C. Traditional herbal remedies used by women in a rural community in northern Maputaland (South Africa) for the treatment of gynaecology and obstetric complaints. *S. Afr. J. Bot.* **2014**, *94*, 129–139. [CrossRef]

31. Maroyi, A. Traditional use of medicinal plants in south-central Zimbabwe: Review and perspectives. *J. Ethnobiol. Ethnomed.* **2013**, *9*, 31. [CrossRef] [PubMed]

32. IUCN. IUCN Red List of Threatened Species (ver. 2009.2). Available online: https://dx.doi.org/10.2305/IUCN. UK.2009-2.RLTS.T168039A6451430.en (accessed on 25 December 2019).

33. SANBI. Red List of South African Plants. Available online: http://redlist.sanbi.org.www57.cpt1.host-h.net/ genus.php?genus=1603 (accessed on 26 December 2019).

34. Ncube, B.; Ndhlala, A.R.; Okem, A.; Van Staden, J. *Hypoxis* (Hypoxidaceae) in African traditional medicine. *J. Ethnopharmacol.* **2013**, *150*, 818–827. [CrossRef]

35. Hammerton, R.D.; Smith, M.T.; Van Staden, J. Factors influencing seed variability and germination in *Hypoxis hemerocallidea* Fisch. & Meyer. *Seed Sci. Technol.* **1989**, *17*, 613–624.

36. Singh, Y. Hypoxis–yellow star of horticulture, folk remedies and conventional medicine. *Veld and Flora* **1999**, *85*, 123–125.

37. Shaik, S.; Govender, K.; Leanya, M. GA$_3$ mediated dormancy alleviation in the reputed African potato, *Hypoxis hemerocallidea*. *Afr. J. Tradit. Complement. Altern. Med.* **2014**, *11*, 330–333. [CrossRef]

38. Hammerton, R.D.; Van Staden, J. Seed germination of *Hypoxis hemerocallidea*. *S. Afr. J. Bot.* **1988**, *54*, 277–280. [CrossRef]

39. Katerere, D.R. Hypoxis hemerocallidea (African potato): A botanical whose time has come? In *African Natural Products: Discoveries and Challenges in Chemistry, Health and Nutrition*; Juliani, H.R., Simon, J.E., Ho, C.T., Eds.; American Chemical Society: Washington, DC, USA, 2013; Volume II, pp. 51–61.

40. Ndong, Y.A.; Wadouachi, A.; Sangwan-Norreel, B.S.; Sangwan, R.S. Efficient in vitro regeneration of fertile plants from corm explants of *Hypoxis hemerocallidea* landrace Gaza–the "African potato". *Plant Cell Rep.* **2006**, *25*, 265–273. [CrossRef] [PubMed]

41. Street, R.A.; Prinsloo, G. Commercially important medicinal plants of South Africa: A review. *J. Chem.* **2013**, 205048. [CrossRef]

42. Boukes, G.J.; Van de Venter, M. *In vitro* modulation of the innate immune response and phagocytosis by three *Hypoxis* spp. and their phytosterols. *S. Afr. J Bot.* **2016**, *102*, 120–126. [CrossRef]

43. Ojewole, J.A.O. Anticonvulsant activity of *Hypoxis hemerocallidea* Fisch. & C.A. Mey. (Hypoxidaceae) corm ('African potato') aqeous extracts in mice. *Phytother. Res.* **2008**, *22*, 91–96. [PubMed]

44. Drewes, S.E.; Khan, F. The African potato (*Hypoxis hemerocallidea*): A chemical-historical perspective. *S. Afr. J. Sci.* **2004**, *100*, 425–430.

45. Koduru, S.; Grierson, D.S.; Afolayan, A.J. Ethnobotanical information of medicinal plants used for treatment of cancer in the Eastern Cape Province, South Africa. *Curr. Sci.* **2007**, *92*, 906–908.

46. Fasinu, P.S.; Gutmann, H.; Schiller, H.; Bouic, P.J.; Rosekranz, B. The potential of *Hypoxis hemerocallidea* for herb-drug interaction. *Pharm. Biol.* **2013**, *51*, 1–9.

47. Mills, E.; Cooper, C.; Seely, D.; Kanfer, I. African herbal medicines in the treatment of HIV: *Hypoxis* and *Sutherlandia*. An overview of evidence and pharmacology. *Nutr. J.* **2005**, *4*, 19–25. [CrossRef]

48. Otunola, G.A.; Afolayan, A.J. Proximate and elemental composition of leaf, corm, root and peel of *Hypoxis hemerocallidea*: A Southern African multipurpose medicinal plant. *Pak. J. Pharm. Sci.* **2019**, *32*, 535–539.

49. De Wet, H.; Nzama, V.N.; Van Vuuren, S.F. Medicinal plants used for the treatment of sexually transmitted infections by lay people in northern Maputaland, KwaZulu–Natal Province, South Africa. *S. Afr. J. Bot.* **2012**, *78*, 12–20. [CrossRef]

50. Semenya, S.S.; Potgieter, M.J.; Erasmus, L.J.C. Indigenous plant species used by Bapedi healers to treat sexually transmitted infections: Their distribution, harvesting, conservation and threats. *S. Afr. J. Bot.* **2013**, *87*, 66–75. [CrossRef]

51. Ojewole, J.A.O.; Olayiwola, G.; Nyinawumuntu, A. Bronchorelaxant property of 'African potato' (*Hypoxis hemerocallidea* corm) aqueous extract in vitro. *J. Smooth Muscle Res.* **2009**, *45*, 241–248. [CrossRef] [PubMed]

52. Nyinawumuntu, A.; Awe, E.O.; Ojewole, J.A.O. Uterolytic effect of *Hypoxis hemerocallidea* Fisch. & C.A. Mey. (Hypoxidaceae) corm ['African potato'] aqueous extract. *J. Smooth Muscle Res.* **2008**, *44*, 167–176. [PubMed]

53. Semenya, S.S.; Maroyi, A. Medicinal plants used for the treatment of tuberculosis by Bapedi traditional healers in three districts of the Limpopo province, South Africa. *Afr. J. Tradit. Complement. Altern. Med.* **2013**, *10*, 316–323. [CrossRef] [PubMed]

54. Barbosa, F.; Hlashwayo, D.; Sevastyanov, V.; Chichava, V.; Mataveia, A.; Boane, E.; Cala, A. medicinal plants sold for treatment of bacterial and parasitic diseases in humans in Maputo city markets, Mozambique. *BCM Compliment. Med. Ther.* **2020**, *20*, 19. [CrossRef]

55. Bruschi, P.; Morganti, M.; Mancini, M.; Sigronini, M.A. Traditional healers and laypeople: A qualitative and quantitative approach to local knowledge on medicinal plants in Muda (Mozambique). *J. Ethnopharmacol.* **2011**, *138*, 543–563. [CrossRef]

56. Viol, D.I.; Chagonda, L.S.; Moyo, S.R.; Mericli, A.H. Toxicity and antiviral activities of some medicinal plants used by traditional medical practitioners in Zimbabwe. *Am. J. Plant Sci.* **2016**, *7*, 1538–1544. [CrossRef]

57. Ncube, B.; Finnie, J.F.; Van Staden, J. Seasonal variation in antimicrobial and phytochemical properties of frequently used medicinal bulbous plants from South Africa. *S. Afr. J. Bot.* **2011**, *77*, 387–396. [CrossRef]

58. Mogatle, S.; Skinner, M.; Mills, E.; Kanfer, I. Effect of African potato (*Hypoxis hemerocallidea*) on the pharmacokinetics of efavirenz. *S. Afr. Med. J.* **2008**, *98*, 945–949.

59. Goboza, M.; Aboua, Y.G.; Meyer, S.; Oguntibeju, O.O. Diabetes mellitus: Economic and health burden, treatment and the therapeutic effects of *Hypoxis hemerrocallidea* plant. *Medical Technol. S. Afr.* **2016**, *40*, 39–46.

60. Nair, V.D.P.; Kanfer, I. Sterols and sterolins in *Hypoxis hemerocallidea* (African potato). *S. Afr. J. Sci.* **2008**, *104*, 322–324.

61. Zimudzi, C. African Potato (*Hypoxis* Spp): Diversity and comparison of the phytochemical profiles and cytotoxicity evaluation of four Zimbabwean Species. *J. Appl. Pharm. Sci.* **2014**, *4*, 079–083.

62. Van Staden, J.; Upfold, S.J. Transport and metabolism of hypoxoside in intact plants of *Hypoxis hemerocallidea*. *S. Afr. J. Bot.* **1994**, *60*, 225–226. [CrossRef]

63. Mkhize, N.; Mohanlall, V.; Odhav, B. Isolation and quantification of β-sitosterol, ergosterol and stigmasterol from *Hypoxis rigidula* Baker var. rigidula and *Hypoxis hemerocallidea* Fisch., C.A.Mey. & Avé-Lall (Hypoxidaceae). *Int. J. Sci.* **2013**, *2*, 118–134.

64. Zibula, S.M.X.; Ojewole, J.A.O. Hypoglycaemic effects of *Hypoxis hemerocallidea* (Fisch. and C.A. Mey.) corm 'African potato' methanolic extract in rats. *Med. J. Islamic World Acad. Sci.* **2000**, *13*, 75–78.

65. Oguntibeju, O.O.; Meyer, S.; Aboua, Y.G.; Goboza, M. *Hypoxis hemerocallidea* significantly reduced hyperglycaemia and hyperglycaemic-induced oxidative stress in the liver and kidney tissues of streptozotocin-induced diabetic male wistar rats. *Evid. Based Complement. Alternat. Med.* **2016**, 1–10. [CrossRef]

66. Liu, Z.; Wilson-Welder, J.H.; Hostetter, J.M.; Jergens, A.E.; Wannemuehler, M.J. Prophylactic treatment with *Hypoxis hemerocallidea* corm (African potato) methanolic extract ameliorates *Brachyspira hyodysenteriae* - induced murine typhlocolitis. *Exp. Biol. Med.* **2010**, *235*, 222–229. [CrossRef]

67. Muwanga, C. An Assessment of *Hypoxis hemerocallidea* Extracts, and Actives as Natural Antibiotic, and Immune Modulation Phytotherapies. Ph.D. Thesis, University of the Western Cape, Cape Town, South Africa, 2006.

68. Cloete, P.C.; Idsardi, E. Bio-fuels and food security in South Africa: The role of indigenous and traditional food crops. In Proceedings of the Selected Paper prepared for presentation at the International Association of Agricultural Economists (IAAE) Triennial Conference, Foz do Iguaçu, Brazil, 18–24 August 2012.

69. Jonnalagadda, S.B.; Kindness, A.; Kubayi, S.; Cele, M.N. Macro, minor and toxic elemental uptake and distribution in *Hypoxis hemerocallidea*, "the African potato"—An edible medicinal plant. *J. Environ. Sci. Health Part B* **2008**, *43*, 271–280. [CrossRef]

70. Wadley, L.; Backwell, L.; d'Errico, F.; Sievers, C. Cooked starchy rhizomes in Africa 170 thousands years ago. *Science* **2020**, *367*, 87–91. [CrossRef]

71. Makunga, N.P.; Philander, L.E.; Smith, M. Current perspectives on an emerging formal natural products sector in South Africa. *J. Ethnopharmacol.* **2008**, *119*, 365–375. [CrossRef]

72. Van Wyk, B.-E. The potential of South African plants in the development of new medicinal products. *S. Afr. J. Bot.* **2011**, *77*, 812–829. [CrossRef]

73. Williams, V.L.; Witkowski, E.T.F.; Balkwill, K. Volume and financial value of species traded in the medicinal plant markets of Gauteng, South Africa. *Int. J. Sust. Dev. World* **2007**, *14*, 584–603. [CrossRef]

74. Ndawonde, B.G.; Zobolo, A.M.; Dlamini, E.T.; Siebert, S.J. A survey of plants sold by traders at Zululand muthi markets, with a view to selecting popular plant species for propagation in communal gardens. *Afr. J. Range Forage Sci.* **2007**, *24*, 103–107. [CrossRef]

75. Dold, A.P.; Cocks, M.L. The trade in medicinal plants in the Eastern Cape Province, South Africa. *S. Afr. J. Sci.* **2002**, *98*, 589–597.

76. Mugomeri, E.; Chatanga, P.; Raditladi, T.; Makara, M.; Tarirai, C. Ethnobotanical study and conservation status of local medicinal plants: Towards a repository and monograph of herbal medicines in Lesotho. *Afr. J. Tradit. Complement. Altern. Med.* **2016**, *13*, 143–156. [CrossRef]

77. Monakisi, C.M. Knowledge and Use of Traditional Medicinal Plants by the Setswana-Speaking Community of Kimberley, Northern Cape of South Africa. Ph.D. Thesis, University of Stellenbosch, Stellenbosch, South Africa, 2007.

78. Krog, M.; Falcão, M.P.; Olsen, C.S. Medicinal plant markets and trade in Maputo, Mozambique. In *Forest and Landscape Working Papers no. 16-2006*; Danish Centre for Forest, Landscape and Planning, KVL Royal Veterinary and Agricultural University: Copenhagen, Denmark, 2006.

79. Kumar, V.; Moyo, M.; Van Staden, J. Somatic embryogenesis in *Hypoxis hemerocallidea*: An important African medicinal plant. *S. Afr. J. Bot.* **2016**, *108*, 331–336. [CrossRef]

80. Gurib-Fakim, A. Medicinal plants: Traditions of yesterday and drugs of tomorrow. *Mol. Aspects Med.* **2006**, *27*, 1–93. [CrossRef]

81. Moyo, M.; Amoo, S.O.; Aremu, A.O.; Gruz, J.; Subrtová, M.; Doležal, K.; Van Staden, J. Plant regeneration and biochemical accumulation of hydroxybenzoic and hydroxycinnamic acid derivatives in *Hypoxis hemerocallidea* organ and callus cultures. *Plant Sci.* **2014**, *227*, 157–164. [CrossRef]

82. George, J.; Laing, M.D.; Drewes, S.E. Phytochemical research in South Africa. *S. Afr. J. Sci.* **2001**, *97*, 93–105.

83. Katerere, D.R.; Eloff, J.N. Anti-bacterial and anti-oxidant activity of *Hypoxis hemerocallidea* (Hypoxidaceae): Can leaves be substituted for corms as a conservation strategy? *S. Afr. J. Bot.* **2008**, *74*, 613–616. [CrossRef]

84. Du Toit, E.S.; Maltzahn, I.V.; Soundy, P. Chemical analysis of cultivated *Hypoxis hemerocallidea* using thin layer chromatography. *HortScience* **2005**, *40*, 1119. [CrossRef]

85. Nigro, S.A.; Makunga, N.P.; Grace, O.M. Medicinal plants at the ethnobotany–biotechnology interface in Africa. *S. Afr. J. Bot.* **2004**, *70*, 89–96. [CrossRef]

86. Okem, A.; Stirk, W.A.; Street, R.A.; Southway, C.; Finnie, J.F.; Van Staden, J. Effects of Cd and Al stress on secondary metabolites, antioxidant and antibacterial activity of *Hypoxis hemerocallidea* Fisch. & C.A. Mey. *Plant Physiol. Biochem.* **2015**, *97*, 147–155. [PubMed]

87. Mofokeng, M.M.; Kleynhans, R.; Sediane, L.M.; Morey, L.; Araya, H.T. Propagation of *Hypoxis hemerocallidea* by inducing corm buds. *S. Afr. J. Plant Soil* **2018**, *35*, 359–365. [CrossRef]

88. Nsibande, B.E.B. In Vitro Regeneration of four *Hypoxis* Species and Transformation of *Camelina sativa* and *Crambe abyssinica*. Ph.D. Thesis, Swedish University of Agricultural Sciences, Uppsala, Sweden, 2012.

89. Nsibande, B.E.B.; Li, X.; Ahlman, A.; Zhu, L.H. In vitro regeneration of endangered medicinal *Hypoxis* species. *Am. J. Plant Sci.* **2015**, *6*, 2585–2595. [CrossRef]

90. Gillmer, M.; Symmonds, R. Seed collection and germination: *Hypoxis hemerocallidea* (Hypoxidaceae). *Plantlife* **1999**, *21*, 36–37.

91. McAlister, B.G.; Van Staden, J. Effect of artificially induced stress conditions on the growth of the medicinal plant *Hypoxis hemerocallidea*. *S. Afr. J. Bot.* **1995**, *61*, 85–89. [CrossRef]

92. Kottek, M.; Grieser, J.; Beck, C.; Rudolf, B.; Rubel, F. World map of the Köppen-Geiger climate classification update. *Meteorol. Z.* **2006**, *15*, 259–263. [CrossRef]

93. White, A.G.; Davies-Coleman, M.T.; Ripley, B.S. Measuring and optimising umckalin concentration in wild-harvested and cultivated *Pelargonium sidoides* (Geraniaceae). *S. Afr. J. Bot.* **2008**, *74*, 260–267. [CrossRef]

94. Kulkarni, M.G.; Street, R.A.; Van Staden, J. Germination and seedling growth requirements for propagation of *Dioscorea dregeana* (Kunth) Dur. and Schinz—A tuberous medicinal plant. *S. Afr. J. Bot.* **2007**, *73*, 131–137. [CrossRef]

95. Mokgehle, S.N.; Tesfay, S.Z.; Araya, H.T.; du Plooy, C.P. Antioxidant activity and soluble sugars of African ginger (*Siphonochilus aethiopicus*) in response to irrigation regimen and nitrogen levels. *Acta Agric. Scand. Sect. B Soil Plant Sci.* **2017**, *67*, 425–434. [CrossRef]
96. Mofokeng, M.M.; Visser, D.; Kleynhans, R.; du Plooy, C.P.; Prinsloo, G.; Soundy, P. Estimation of *Pelargonium sidoides* root damage by *Meloidogyne* spp. *J. Entomol. Nematol.* **2013**, *5*, 38–41. [CrossRef]

Article

Diversity and Structure of an Arid Woodland in Southwest Angola, with Comparison to the Wider Miombo Ecoregion

John L. Godlee [1,*], Francisco Maiato Gonçalves [2], José João Tchamba [2], Antonio Valter Chisingui [2], Jonathan Ilunga Muledi [3], Mylor Ngoy Shutcha [3], Casey M. Ryan [1], Thom K. Brade [1] and Kyle G. Dexter [1,4]

[1] School of GeoSciences, University of Edinburgh, Edinburgh EH9 3FF, UK
[2] Herbarium of Lubango, ISCED Huíla, Sarmento Rodrigues Str. No. 2, CP. 230, Lubango, Angola
[3] Ecologie, Restauration Ecologique et Paysage, Faculté des Sciences Agronomique, Université de Lubumbashi, Route Kasapa BP 1825, Congo
[4] Royal Botanic Garden Edinburgh, Edinburgh EH3 5LR, UK
[*] Correspondence: johngodlee@gmail.com

Received: 16 March 2020; Accepted: 31 March 2020; Published: 3 April 2020

Abstract: Seasonally dry woodlands are the dominant land cover across southern Africa. They are biodiverse, structurally complex, and important for ecosystem service provision. Species composition and structure vary across the region producing a diverse array of woodland types. The woodlands of the Huíla plateau in southwest Angola represent the extreme southwestern extent of the miombo ecoregion and are markedly drier than other woodlands within this ecoregion. They remain understudied, however, compared to woodlands further east in the miombo ecoregion. We aimed to elucidate further the tree diversity found within southwestern Angolan woodlands by conducting a plot-based study in Bicuar National Park, comparing tree species composition and woodland structure with similar plots in Tanzania, Mozambique, and the Democratic Republic of Congo. We found Bicuar National Park had comparatively low tree species diversity, but contained 27 tree species not found in other plots. Plots in Bicuar had low basal area, excepting plots dominated by *Baikiaea plurijuga*. In a comparison of plots in intact vegetation with areas previously disturbed by shifting-cultivation agriculture, we found species diversity was marginally higher in disturbed plots. Bicuar National Park remains an important woodland refuge in Angola, with an uncommon mosaic of woodland types within a small area. While we highlight wide variation in species composition and woodland structure across the miombo ecoregion, plot-based studies with more dense sampling across the ecoregion are clearly needed to more broadly understand regional variation in vegetation diversity, composition and structure.

Keywords: woodland; miombo; savanna; diversity; disturbance; Baikiaea

---

1. Introduction

Tropical woodlands extend over 12 countries in central and southern Africa, with an estimated area of ~3.7 million km^2 [1–3]. Within this, miombo woodlands are the dominant vegetation type, characterised by trees of the *Brachystegia*, *Julbernardia* and *Isoberlinia* genera, all within the Fabaceae family, subfamily Detaroideae [4–6]. These genera are seldom found as dominant species outside miombo woodlands, and while their contribution to the biomass of miombo woodlands is substantial, it varies throughout the region [5]. Across the range of southern African woodlands, variation in climate, edaphic factors, disturbance regimes and biogeography maintain a diverse array of woodland types in terms of both species composition and physiognomy [7–9]. Many of these woodlands have a flammable grassy understorey and thus are also considered to be a form of savanna [10].

The miombo ecoregion extends across the continent in a wide band that reaches north into Kenya and the Democratic Republic of Congo (DRC) and south into the northeast of South Africa (Figure 1a). Miombo woodlands are defined both by their tree diversity and by their structure of a grassy herbaceous understorey with an often sparse tree canopy. In archetypical miombo woodlands, species of the genera *Brachystegia*, *Julbernardia* and *Isoberlinia* generally hold the most biomass, forming a mostly open woodland canopy. Distinct from dry tropical forests, miombo woodlands generally maintain a grassy understorey dominated by grass species using the C_4 carbon fixation pathway [11]. Miombo woodlands are heavily structured by seasonal fire and herbivory, with fire particularly often preventing the creation of a closed tree canopy which would naturally occur in the absence of these disturbances [12,13]. Within the miombo ecoregion, other woodland types exist, notably, woodlands dominated by *Baikiaea plurijuga* or *Colophospermum mopane* [5].

Southern African woodlands are structurally complex but species poor in the tree layer compared to dry tropical forests which exist at similar latitudes [14,15]. These woodlands contain many endemic tree species, however, and support a highly diverse woodland understorey, with an estimated 8500 species of vascular plants [16]. Miombo woodlands provide ecosystem services for an estimated 150 million people [17]. Additionally, miombo woodlands hold ~18–24 Pg C in woody biomass and soil organic carbon, which is comparable to that held in the rainforests of the Congo basin (~30 Pg C) [18]. As woodland resource extraction and conversion to agricultural land accelerates due to growing human populations, the conservation of miombo woodlands as a biodiverse and unique ecosystem has become a growing concern. Despite their importance however, dry tropical woodlands remain understudied compared to wet forests across the globe [19].

Over the previous two decades, the limited ecological research in southern African woodlands has been concentrated in the central and eastern parts of the miombo region, notably in southern Tanzania, Mozambique, Malawi, Zimbabwe and Zambia. The southwestern extent of miombo woodlands, which is found entirely within Angola has received considerably less attention [20]. Partly this is due to diminished research capacity during the Angolan civil war following the country's independence, which took place officially between 1975 and 2002, but with sporadic localised periods of civil unrest until around 2012 [21]. While botanical surveys of woodlands in this region are more plentiful [20,22], joint studies of woodland species composition and physical structure remain scarce. This is despite the value of these studies in helping to estimate woodland net primary productivity, carbon sequestration potential, and studies of community assembly. To properly understand spatial variation in woodland species composition and physical structure across the miombo ecoregion, it is necessary to fill understudied gaps. In this study, we aim to address one such gap in southwest Angola, and place it in context with other woodlands across the miombo ecoregion.

The miombo woodlands of southwest Angola are found in their most intact form in Bicuar National Park and to a lesser extent in the adjacent Mupa National Park, on the Huíla plateau [23]. Both of these national parks have been protected to varying extents since 1938 [20]. These woodlands exist in much drier conditions than other miombo woodlands, precipitation diminishes rapidly within the Huíla plateau towards the Angolan coast and the Namib desert (Figure 1a). The vegetation of the Huíla plateau holds many endemic species, around 83 endemic Fabaceae species [24] and the most endemic plant species of any part of Angola [25]. Linder [26] and Droissart et al. [27] both identify the western portion of the Huíla plateau as a centre of tropical African endemism.

Much of the historic miombo woodland area in southwest Angola surrounding the Bicuar and Mupa National Parks has been deforested in recent years, with a clear increase in deforestation activity since the end of the civil war owing to an increase in rural population and agricultural activity [20,28]. The western extent of miombo woodlands found within Bicuar National Park are therefore of great importance for conservation as a refuge for wildlife and endemic plant species [20].

It is important to focus not only on the biodiversity of undisturbed woodland areas but also previously disturbed land in order to properly assess the biodiversity and woodland structure of the Park. Woodland disturbance through shifting cultivation practices produces novel habitats which are not necessarily of lower conservation value [29,30]. Since Bicuar National Park's rejuvenation following the reinforcement of park boundaries after the civil war, many areas of woodland that were previously heavily grazed, farmed via shifting cultivation techniques, and used for timber extraction have been allowed to re-establish and are now protected from further human resource extraction. This presents a unique opportunity to compare the species composition of these disturbed areas with areas of nearby woodland that have not been farmed in living memory.

In this study, we present results of the tree diversity and woodland structure of miombo woodlands found at the far western extent of miombo woodlands in Bicuar National Park, Huíla province, Angola. Our study utilised recently installed biodiversity monitoring plots set up within the Park in 2018 and 2019. We compare the tree diversity and woodland structure of Bicuar National Park with biodiversity monitoring plots previously established in other areas of miombo woodland across the miombo ecoregion which use a common plot biodiversity census methodology. In addition, we take advantage of a unique opportunity to compare the tree species composition of areas of abandoned and now protected farmland that have begun to re-establish as woodland. Specifically, this study aims to:

1. Describe the tree species diversity and structure of woodlands in Bicuar National Park, and compare this composition with other woodlands across the miombo eco-region.
2. Explore the role of environmental factors in driving changes in tree species composition across the miombo ecoregion.
3. Describe variation in tree species composition and woodland structure between disturbed and undisturbed woodland patches within Bicuar National Park.

2. Materials and Methods

2.1. Study Area

We chose three areas of miombo woodland across the miombo ecoregion to compare with those in Bicuar National Park, Angola (S15.1°, E14.8°). The three sites were Gorongosa District in central Mozambique (S19.0°, E34.2°) [31], Kilwa District in southern Tanzania (S9.0°, E39.0°) [32], and the Mikembo Natural Reserve in Katanga, southern Democratic Republic of Congo (DRC) (S11.5°, E27.7°) [33]. Within each of these woodland sites, multiple one hectare square plots had been installed previously to monitor biodiversity and biomass dynamics. In Katanga, a larger 10 ha plot was subdivided into ten 1 ha plots for this study. We used these previous censuses, collected between 2010 and 2019, to estimate tree biodiversity and woodland structure. Sites range in Mean Annual Precipitation (MAP) from 864 mm y^{-1} in Bicuar to 1115 mm y^{-1} in Katanga. Mean Annual Temperature ranges from ~20.5 °C in Bicuar and Katanga to ~25.8 °C in Kilwa (Figure 1b, Table 1).

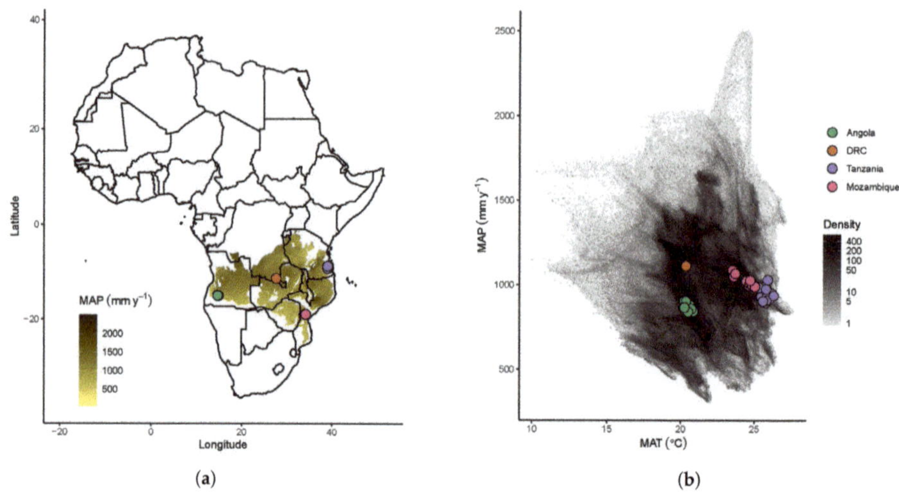

Figure 1. Locations of plots used in this study, by (**a**) geographic location with respect to the distribution of miombo woodland vegetation (shaded brown according to mean annual precipitation) [1], and (**b**) showing the plot locations compared to the climate space of the miombo ecoregion estimated using the WorldClim dataset over the miombo woodland vegetation extent with a pixel size of 30 arc seconds (0.86 km² at the equator) [34]. Please note that the density colour scale is log-transformed for visual clarity.

Table 1. Description of each group of plots used in the analysis. MAT = Mean Annual Temperature, MAP = Mean Annual Precipitation, CWD = Climatic Water Deficit, DD = Decimal Degrees.

Plot Group	MAT (°C)	MAP (mm y⁻¹)	CWD (mm y⁻¹)	Latitude (DD)	Longitude (DD)	N Plots	N Species
Bicuar NP	20.5	864	−815	−15.12	14.81	15	49
DRC	20.4	1115	−762	−11.49	27.67	12	89
Mozambique	24.4	1029	−662	−18.95	34.16	15	162
Tanzania	25.8	956	−754	−9.05	39.05	22	248

Bicuar National Park covers an area of ~7900 km², established as a hunting reserve in 1938, and later as a national park in 1964 (Figure 2). While fauna populations in the Park were severely damaged by the Angolan civil war, the interior of the Park remains as a largely intact mosaic of miombo woodland, Baikiaea-Burkea woodland, shrub/thicket vegetation and seasonally flooded grassland. Encroachment of agriculture and grazing, particularly along the northwest and western boundaries of the Park, has led to a fragmented park boundary with patches of diminished thicket and woodland in areas of previously farmed land that have been protected since park boundaries were re-established following the end of the civil war.

Plots in Tanzania were located predominantly within or near the Mtarure Forest Reserve, administrated by the Tanzania Forest Service and protected from human incursion since their installation. Plots were established between 2010 and 2011 in grassy savanna/woodland areas, with plots located along the road network with a 1 km buffer from the road. Plots in Mozambique were established in 2004, in areas of miombo woodland that had been previously used for agriculture but since left fallow, and areas of undisturbed miombo woodland, located along the road network, with all plots >250 m from the road. Plots in DRC were established in 2009 and located within a larger 800 ha miombo woodland reserve, which consists of undisturbed miombo woodlands. All plots were located quasi-randomly, with consideration to accessibility for future woodland censuses.

2.2. Plot Data Collection

We sampled 15 one hectare plots in Bicuar National Park and collated data from a total of 64 one hectare plots across the miombo ecoregion within four sites. Figure 1a and Table 1 show the locations and general description of each site, respectively. Plots in Bicuar were situated at least 500 m from the edge of a woodland patch to prevent edge effects which may have altered tree species composition.

Within each plot, every tree stem ≥5 cm stem diameter was recorded, except in the DRC plots, where only stems ≥10 cm stem diameter were recorded. For each tree stem the species and stem diameter were recorded. Tree species were identified using local botanists at each site and taxonomy was later checked against the African Plant Database [35]. At all sites, we used Palgrave [36], along with other texts, to identify tree species. Specimens that could not be identified in the field, or subsequently at herbaria, were described as morphospecies. All tree species within the Bicuar National Park plots were identified. Tree coppicing due to fire, herbivory, and human actions is common in miombo woodlands, therefore, for trees with multiple stems, each stem ≥5 cm stem diameter was recorded, while the parent tree was also recorded for diversity analyses described below.

Stem diameter was recorded at 1.3 m from the ground along the stem (diameter at breast height, DBH) as per convention using a diameter tape measure [37]. Where stem abnormalities were present at 1.3 m from the ground, which precluded the accurate estimation of stem diameter at 1.3 m, the stem diameter was recorded at the nearest 10 cm increment above 1.3 m without significant stem abnormalities [37]. To ensure consistency among stem diameter values recorded at different heights, when the stem diameter was recorded at a height other than 1.3 m the stem diameter at 1.3 m was estimated from the recorded stem diameter using a cubic polynomial equation which adjusts for tree stem taper. This equation was calibrated on 100 stems measured at multiple heights in Niassa Province, Mozambique (Appendix A). Stems below 10 cm stem diameter were not measured in the DRC plots. We therefore estimated the number of 5–10 cm stems in each these plots by extrapolating a linear regression of log stem abundance across the available stem diameter classes.

In addition to the one hectare plots across the miombo ecoregion, we compared the tree biodiversity of undisturbed areas of miombo woodland in Bicuar National Park with areas of disturbed woodland around the edge of the Park that had been previously farmed via shifting cultivation methods, and had since been abandoned and reclaimed within the Park boundaries (Figure 2). We identified areas previously farmed with the help of park rangers and local residents who identified these areas from memory. We conducted 20 plot surveys of woodland diversity and structure in these areas with 20 × 50 m (0.1 ha) plots, and compared their diversity and structure with 20 × 50 m subsamples of the 15 one hectare plots within the Park interior. Like the one hectare plots, within these smaller 20 × 50 m plots we recorded the species and stem diameter of every tree stem ≥5 cm stem diameter.

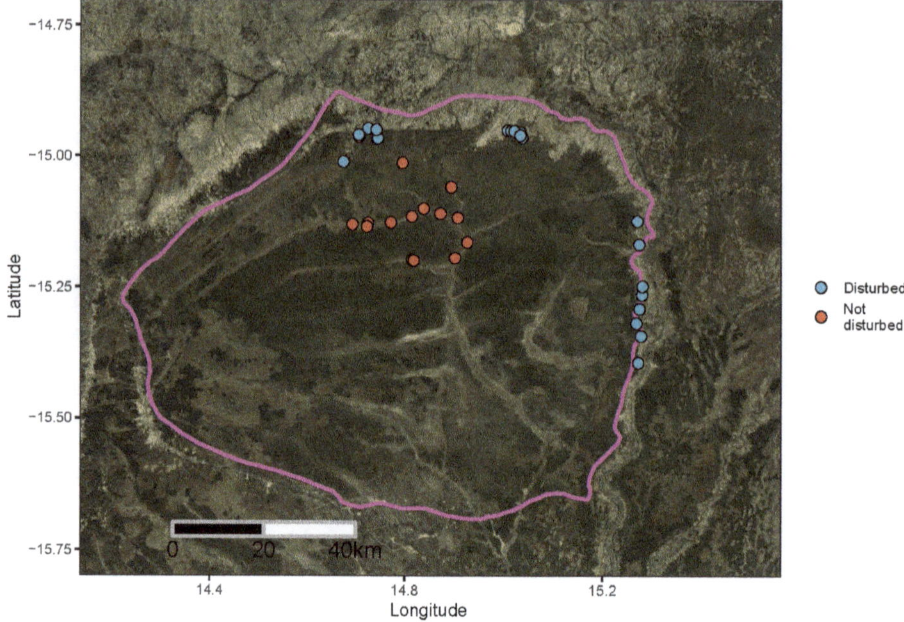

Figure 2. Location of plots in Bicuar National Park, southwest Angola. The Park boundary is shown as a pink outline, according to UNEP-WCMC and IUCN [38]. One hectare undisturbed plots are shown as red points, while disturbed 20 × 50 m (0.1 hectare) plots are shown as blue points. The map background is a true colour composite satellite image generated using the Google Maps Static Maps API in the ggmap R package [39].

2.3. Climatic Data

The WorldClim dataset [34] was used to gather data on plot-level climatic conditions. We estimated Mean Annual Precipitation (MAP) as the mean of total annual precipitation values between 1970 and 2000, and Mean Annual Temperature (MAT) as the mean of mean annual temperatures between 1970 and 2000. The seasonality of temperature (MAT SD) was calculated as the standard deviation of monthly temperature per year, respectively. We estimated Climatic Water Deficit (CWD) for each plot according to Chave et al. [40], as the sum of the difference between monthly rainfall and monthly evapotranspiration when the difference is negative, using the dataset available at http://ups-tlse.fr/pantropical_allometry.htm, which uses data from the WorldClim dataset 1970–2000.

2.4. Data Analysis

We calculated the basal area of each stem (g_i) using:

$$g_i = \pi \times (d_i/2)^2$$

(1)

where d_i is the estimated stem diameter of stem i at 1.3 m having accounted for tree taper. We then calculated the total basal area of each plot as the sum of each stem's basal area. For the DRC plots which lacked 5–10 cm stems, we estimated basal area in this stem diameter class from our extrapolation of stem abundance in the 5–10 cm diameter class, assuming a mean stem diameter of 7.5 cm.

All diversity measures were calculated on individual tree-level data, rather than stem-level data, to avoid artificial inflation of abundance for those species which readily coppice. We calculated the alpha diversity of each plot using both the tree species richness of trees with stems $\geq$5 cm diameter, and the Shannon-Wiener index (H') (Equation (2)), using the vegan package in R [41]:

$$H' = -\sum_{i=1}^{S} p_i \ln p_i \qquad (2)$$

where S is the total number of species in the plot, p_i is the proportional abundance of the ith species and ln is the natural logarithm.

We calculated the pairwise beta diversity among sites using the Sørensen coefficient (S_S) (Equation (3)) [42]:

$$S_S = \frac{2a}{2a + b + c} \qquad (3)$$

where a is the number of species shared between two sites, b is the number of species unique to site 1 and c is the number of species unique to site 2. We calculated S_S for each pairwise combination of sites using aggregated species composition data from all plots in each site. The value of S_S, which ranges between zero and one, was multiplied by 100 to give a "percentage similarity" between communities in species composition.

We estimated abundance evenness for each plot using the Shannon equitability index ($E_{H'}$) [43] which is the ratio of H' to the log transformed species richness.

We analysed the difference in alpha diversity measures and woodland structural variables among groups of plots using Analysis of Variance (ANOVA) statistical models, with a null hypothesis that there was no difference among the mean values of groups of plots. Post-hoc Tukey's HSD tests were used to investigate the degree to which pairwise combinations of plot groups differed in each case.

We used Non-metric Multidimensional Scaling (NMDS) to assess the variation in species composition among one hectare plots, and also between disturbed and undisturbed 20 $\times$ 50 m plots within Bicuar National Park, using the vegan R package. The number of dimensions for NMDS was minimised while ensuring the stress value of the NMDS fit was $\leq$0.1. NMDS analyses were run with 500 random restarts to ensure a global solution was reached. We used Bray-Curtis dissimilarity as the optimal measure of ecological distance [44]. We fit plot-level estimates of MAP, MAT, the seasonality of MAT and CWD to the first two axes of the resulting ordination using the envfit function in the vegan R package to investigate how these environmental factors influenced the grouping of species composition among plots. All analyses were conducted in R version 3.6.1 [45].

3. Results

3.1. Alpha Diversity

In Bicuar National Park we measured a total of 6565 trees within the one hectare plots, and across the four sites, a total of 25,525 trees were sampled. Trees in Bicuar National Park belonged to 48 species within 18 families. Across all four sites we recorded 468 species from 43 families. The most diverse family within each site and among all plots was Fabaceae with 61 species. We encountered 27 tree species in Bicuar National Park which were not found in the other miombo woodland plots (Table 2). The most common of these unique species were *Brachystegia tamarindoides* (n = 576), *Baikiaea plurijuga* (n = 331) and *Baphia massaiensis* (n = 303). Four species unique to Bicuar National Park within this dataset only had one individual recorded: *Elachyptera parvifolia*, *Entandrophragma spicatum*, *Oldfieldia dactylophylla*, *Peltophorum africanum*.

Table 2. Species found in one hectare plots in Bicuar National Park. Stem diameter and basal area are the mean of all stems with the standard error of the mean in parentheses. Number of stems per hectare is the mean of the number of stems in all one hectare plots where stems of that species are present with the standard error of the mean in parentheses. Species found only in Bicuar National Park are marked in bold text with an asterisk.

Family	Species	Stem Diam. (cm)	Basal Area ($m^2\ ha^{-1}$)	N Stems	N Stems ha^{-1}
Fabaceae	Albizia antunesiana	9.1(2.03)	0.07(0.040)	40	8(4.81)
Fabaceae	*Baikiaea plurijuga	28.9(0.75)	1.72(0.570)	331	55.2(17.83)
Fabaceae	*Baphia bequaertii	7.4(0.36)	0.08(0.050)	127	31.8(18.14)
Fabaceae	*Baphia massaiensis	6.6(0.17)	0.05(0.020)	303	30.3(11.20)
Fabaceae	Bobgunnia madagascariensis	7.8(0.91)	0.04(0.020)	32	10.7(9.67)
Fabaceae	*Brachystegia longifolia	12.9(0.48)	1.14(0.430)	576	115.2(72.67)
Fabaceae	Brachystegia spiciformis	11.4(0.52)	0.74(0.430)	326	81.5(46.56)
Phyllanthaceae	*Bridelia mollis	5.7(0.31)	0.02(NA)	23	23(NA)
Fabaceae	Burkea africana	8.5(0.33)	0.39(0.120)	863	71.9(19.11)
Combretaceae	Combretum apiculatum	7.6(0.45)	0.06(0.040)	60	30(15.00)
Combretaceae	Combretum celastroides	5.6(0.34)	<0.01(0.000)	7	3.5(2.50)
Combretaceae	Combretum collinum	6.3(0.09)	0.07(0.020)	609	50.8(20.48)
Combretaceae	*Combretum hereroense	6.7(0.26)	0.02(0.010)	73	12.2(5.69)
Combretaceae	*Combretum psidioides	7.4(0.43)	0.01(0.010)	33	6.6(4.17)
Combretaceae	Combretum zeyheri	6.3(0.35)	0.01(0.000)	61	10.2(3.03)
Euphorbiaceae	*Croton gratissimus	6.1(1.55)	<0.01(NA)	4	4(NA)
Ebenaceae	*Diospyros batocana	8.4(2.14)	<0.01(0.000)	2	1(0.00)
Ebenaceae	*Diospyros kirkii	9.3(1.64)	0.03(NA)	11	11(NA)
Apocynaceae	Diplorhynchus condylocarpon	8.2(0.52)	0.08(0.060)	174	19.3(7.57)
Malvaceae	*Dombeya rotundifolia	5.5(0.19)	<0.01(NA)	2	2(NA)
Celastraceae	*Elachyptera parvifolia	7.3(NA)	<0.01(NA)	1	1(NA)
Meliaceae	*Entandrophragma spicatum	14.6(NA)	<0.01(NA)	1	1(NA)
Fabaceae	Erythrophleum africanum	9.0(0.84)	0.10(0.040)	128	18.3(6.82)
Rubiaceae	*Gardenia volkensii	5.6(1.15)	<0.01(0.000)	5	2.5(1.50)
Fabaceae	*Guibourtia coleosperma	7.2(1.00)	0.02(0.010)	31	6.2(3.54)
Phyllanthaceae	Hymenocardia acida	5.9(1.25)	<0.01(NA)	6	6(NA)
Fabaceae	Julbernardia paniculata	10.1(0.21)	0.92(0.200)	1624	162.4(50.60)
Fabaceae	*Lonchocarpus nelsii	13.4(0.88)	0.15(0.030)	165	15(2.77)
Dipterocarpaceae	*Monotes angolensis	7.4(0.83)	<0.01(0.000)	2	1(0.00)
Ochnaceae	*Ochna pulchra	6.5(0.80)	0.01(0.000)	26	8.7(3.76)
Picrodendraceae	*Oldfieldia dactylophylla	8.5(NA)	<0.01(NA)	1	1(NA)
Fabaceae	*Peltophorum africanum	11.5(NA)	<0.01(NA)	1	1(NA)
Fabaceae	Pericopsis angolensis	8.4(0.61)	0.06(0.020)	97	12.1(5.08)
Phyllanthaceae	Pseudolachnostylis maprouneifolia	6.7(0.45)	0.03(0.010)	84	9.3(3.00)
Combretaceae	*Pteleopsis anisoptera	6.8(0.46)	0.07(0.020)	81	20.2(15.11)
Fabaceae	Pterocarpus angolensis	13.0(0.61)	0.15(0.100)	102	17(8.65)
Fabaceae	*Pterocarpus lucens	6.9(0.94)	<0.01(NA)	4	4(NA)
Rubiaceae	*Rothmannia engleriana	6.8(0.66)	<0.01(0.000)	5	1.7(0.67)
Euphorbiaceae	*Schinziophyton rautanenii	8.0(2.82)	<0.01(NA)	3	3(NA)
Polygalaceae	Securidaca longepedunculata	7.3(1.12)	<0.01(0.010)	4	2(1.00)
Loganiaceae	Strychnos cocculoides	10.4(1.17)	0.03(0.020)	19	6.3(3.53)
Loganiaceae	*Strychnos pungens	6.1(0.48)	<0.01(0.000)	18	3.6(0.93)
Loganiaceae	Strychnos spinosa	6.8(0.36)	0.02(0.010)	97	9.7(4.07)
Combretaceae	*Terminalia brachystemma	6.5(0.21)	0.04(0.020)	174	29(12.04)
Combretaceae	Terminalia sericea	7.1(0.28)	0.06(0.030)	214	23.8(12.18)
Ximeniaceae	Ximenia americana	6.1(0.53)	<0.01(0.000)	7	1.8(0.25)
Sapindaceae	Zanha africana	9.4(1.12)	0.01(NA)	6	6(NA)
Rhamnaceae	*Ziziphus abyssinica	5.9(1.13)	<0.01(NA)	2	2(NA)

Alpha diversity in Bicuar National Park was low compared to other sites (Figure 3). Mean H' across plots in Bicuar National Park was 1.6 ± 0.13. An ANOVA showed a significant difference in H' among sites (F(3,60) = 7.54, $p < 0.01$, Table 3), and a post-hoc Tukey's test showed that H' in plots in Bicuar National Park was significantly different from those in DRC ($H' = 2.7 \pm 0.19, p < 0.01$), Mozambique ($H' = 2.4 \pm 0.2, p < 0.01$) and Tanzania ($H' = 2.2 \pm 0.11, p < 0.05$). Variation in H' is large within Bicuar National Park, with H' ranging from 0.85 to 2.56, but this was a similar range to other sites. In contrast, the range of species richness within Bicuar National Park was much lower than other sites, suggesting that the wide range in H' was caused by variation in abundance evenness.

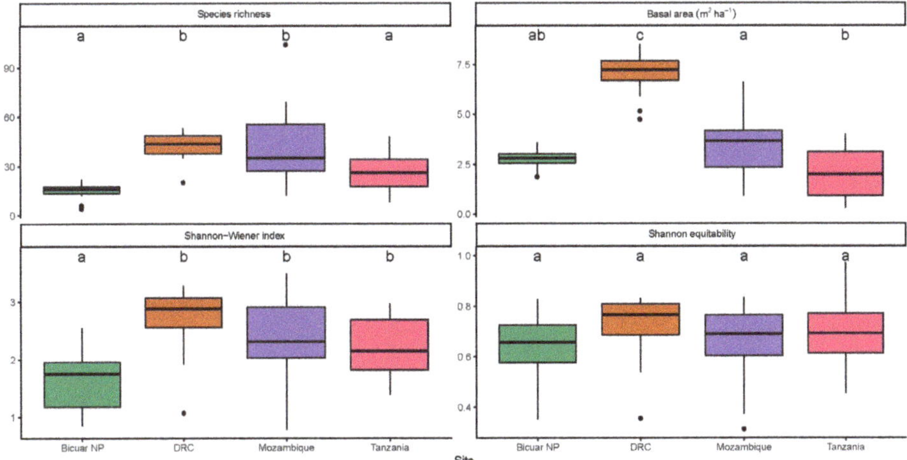

Figure 3. Variation of alpha diversity estimates and basal area among sites. Boxes bound the 1st and 3rd quartiles, with the median within the box. Whiskers represent 1.5 times the interquartile range plus or minus the 1st and 3rd quartiles, respectively. Values found beyond the whiskers are shown individually as points. Letter labels above each box refer to groupings from post-hoc Tukey's tests on the ANOVA of each diversity/structure variable. Sites sharing a letter do not differ significantly ($p < 0.05$).

Table 3. Results of ANOVA tests for alpha diversity metrics and plot basal area, among the four sites. Mean values for each site with standard errors in parentheses are shown. Asterisks indicate the *p*-value of individual sites in each ANOVA (*** <0.001, ** <0.01, * <0.05, . <0.1).

	Dependent Variable:			
	Species Richness (1)	**Basal Area** (2)	**Shannon (H')** (3)	**Shannon Equit. (E_H)** (4)
DRC	27.920 ***	4.175 ***	1.055 ***	0.080
	(5.538)	(0.452)	(0.236)	(0.053)
Tanzania	12.440 **	−0.721 *	0.605 ***	0.064
	(4.788)	(0.391)	(0.204)	(0.046)
Mozambique	27.930 ***	0.653	0.792 ***	0.028
	(5.221)	(0.427)	(0.223)	(0.050)
Constant	14.330 ***	2.778 ***	1.617 ***	0.631 ***
	(3.692)	(0.302)	(0.158)	(0.035)
Observations	64	64	64	64
Adjusted R^2	0.363	0.691	0.237	0.003
Residual Std. Error (df = 60)	14.300	1.168	0.611	0.137
F Statistic (df = 3; 60)	12.980 ***	48.040 ***	7.537 ***	1.000

3.2. Beta Diversity

The NMDS of plot species composition among one hectare plots was run with four dimensions. The stress value was 0.10. Plot diversity in Bicuar National Park formed three distinct groups within axes 1 and 2 of the NMDS ordination. Bicuar plots 9, 13 and 15 were characterised by high abundances of *Baikiaea plurijuga*, *Baphia massaiensis* and *Croton gratissimus*, according to species scores from the

NMDS. Bicuar plots 4, 11 and 12 were characterised by *Brachystegia tamarindoides* and *Ochna pulchra*. The third group consisting of the remaining seven plots surprisingly had a species composition most similar to that of plots in the DRC group according to the NMDS, sharing the core miombo species of *Julbernardia paniculata* and *Pterocarpus angolensis*. This group of plots in Bicuar National Park was further characterised by the abundance of *Pterocarpus lucens*, *Strychnos pungens* and *Bridelia mollis* however, which were not present in the DRC plots. All environmental factors fitted to the NMDS ordination correlated significantly with the grouping of plots (Figure 4a). MAT explained the most variation in plot position on the first two NMDS axes ($R^2 = 0.75$, $p < 0.01$), followed by CWD ($R^2 = 0.54$, $p < 0.01$), the seasonality of MAT ($R^2 = 0.46$, $p < 0.01$) and MAP ($R^2 = 0.4$, $p < 0.01$). Variation in MAP explained much of the difference among plots in Bicuar National Park versus those in Tanzania and Mozambique. Axes 3 and 4 showed a greater degree of overlap in species composition among plot groups, with plots from Bicuar National Park similar to a select few plots in both Tanzania and Mozambique (Figure 4b). Axis 3 distinguished plots in Bicuar NP from those in DRC, while plots from all geographic area overlapped in their distribution across Axis 4. Axes 3 and 4 largely reflected distribution patterns of less abundant species and not the dominant species in the vegetation.

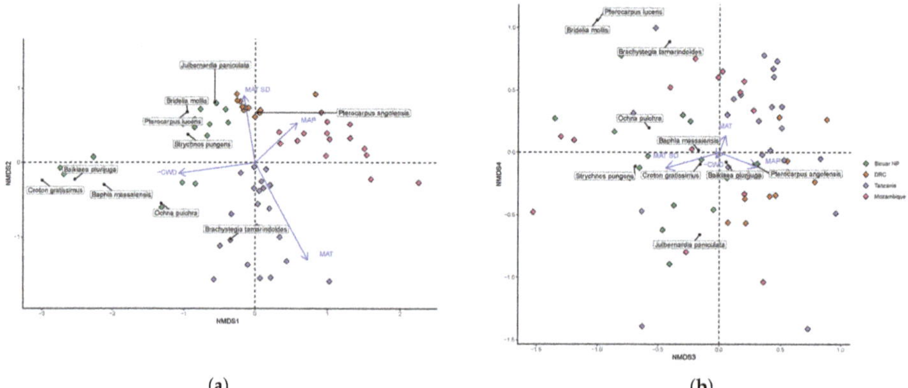

(a)   (b)

Figure 4. Environmental factors fitted to axes 1 and 2 (**a**), 3 and 4 (**b**) of the NMDS ordination of species composition of one hectare plots, showing the variation in plot species composition within and among sites. Diamonds are plot scores coloured by site. The lengths of arrows indicating environmental factor fits are scaled by R^2. Arrows point in the direction of increasing values of that environmental factor. Please note that Climatic Water Deficit (CWD) is expressed more intuitively as the negative inverse of CWD, thus larger values indicate higher levels of CWD.

The pairwise Sørensen coefficient of percentage similarity (S_S) showed that the species composition of plots in Bicuar National Park had low similarity with other sites in the study, sharing a few species with other sites (Table 4). Similar to the NMDS, these results show that plots in Bicuar National Park are most similar to those found in DRC.

Table 4. Pairwise beta diversity comparison of plot groups measured by the Sørensen coefficient (S_S) of percentage similarity of aggregated plot level data from each of the four sites. Values in parentheses are the number of species unique to each site in each comparison.

Site 1	Site 2	S_S	Shared Species
Bicuar NP (34)	DRC (74)	20.6	14
Bicuar NP (34)	Tanzania (147)	13.4	14
Bicuar NP (37)	Mozambique (236)	7.5	11
DRC (64)	Tanzania (137)	19.3	24
DRC (69)	Mozambique (228)	11.3	19
Tanzania (139)	Mozambique (225)	10.8	22

3.3. Woodland Structure

Mean basal area of plots in Bicuar National Park was 2.78 ± 0.122 m^2 ha^{-1}, ranging from 1.86 to 8.53 m^2 ha^{-1} (Figure 3). An ANOVA showed a significant difference in basal area among sites (F(3,60) = 48.04, $p < 0.01$), and a post-hoc Tukey's test showed that basal area in Bicuar National Park was significantly lower than plots in DRC (BA = 6.95 ± 0.327 m^2 ha^{-1}, $p < 0.01$), but there were no significant differences between Bicuar and Mozambique (BA = 3.43 ± 0.409 m^2 ha^{-1}, $p = 0.43$) or Tanzania (BA = 2.06 ± 0.253 m^2 ha^{-1}, $p = 0.26$) (Figure 3). Additionally, Bicuar plots had less variation in basal area among plots than other sites. Plots in Bicuar with the highest basal area were dominated by *Baikiaea plurijuga* and *Baphia massaiensis* (Plots 9, 13, and 15).

The stem diameter abundance distribution in Bicuar National Park was comparable with other sites (Figure 5), albeit with fewer stems in each class. The slope of log mean stem size distribution among diameter bins was -0.92 ± 0.067 in Bicuar National Park, -0.99 ± 0.067 in DRC, -0.89 ± 0.065 in Tanzania, and -0.87 ± 0.075 in Mozambique.

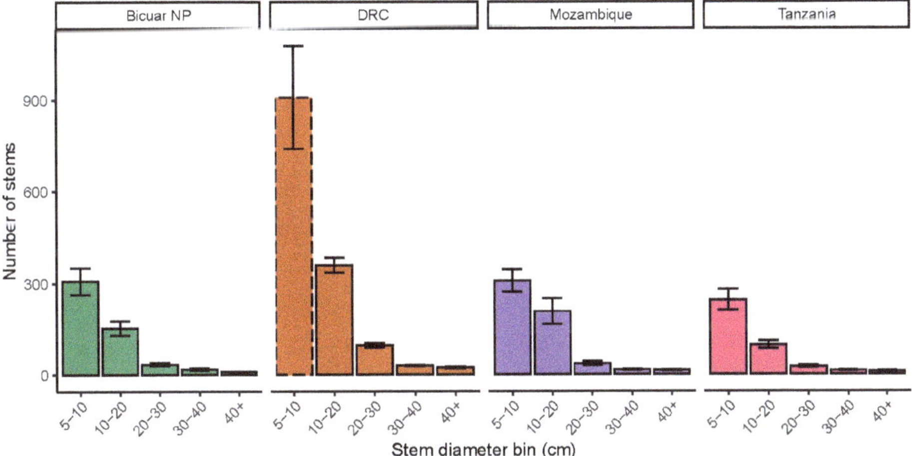

Figure 5. Ranked variation between plots in stem number within each site, with bars according to stem diameter class. Error bars are the mean $\pm$ 1 standard error. The dashed bar for the DRC 5–10 cm stem diameter class indicates that these measurements were estimated by extrapolating a linear regression of log stem abundance across the available stem diameter classes for DRC.

3.4. Effect of Disturbance via Shifting Cultivation on Diversity within Bicuar National Park

There was a clear difference in the species composition of previously farmed disturbed woodland plots and undisturbed woodland plots, but with some overlap (Figure 6). Notably, Plots 4 and 7 in putatively undisturbed woodland have a species composition more resembling the disturbed plots.

These two plots were dominated by *Brachystegia tamarindoides* and *Burkea africana*, with *B. africana* being a species which occurred frequently as a pioneer in the disturbed plots. The undisturbed plots 15, 13 and 9 represent distinct outliers in the NMDS. These three plots were dominated by *Baikiaea plurijuga* which was not encountered in the disturbed plots. The most common species in the disturbed plots was *Baphia massaiensis* ($n = 158$), with a mean stem diameter of 6.1 ± 1.87 cm, while in the undisturbed plots the most common species was *Julbernardia paniculata* ($n = 125$), with a mean stem diameter of 11.8 ± 7.24 cm. Mean alpha diversity was marginally higher in disturbed plots ($H' = 1.7 \pm 0.08$) than in undisturbed plots ($H' = 1.3 \pm 0.14$) and an ANOVA showed that there was a significant difference in H' between the two plot types ($F(1,33) = 5.91$, $p < 0.05$) (Figure 7, Table 5). Mean plot species richness was also lower in undisturbed plots (6.4 ± 0.86) than disturbed plots (8.7 ± 0.53). Mean $E_{H'}$ was 0.8 ± 0 in disturbed plots and 0.7 ± 0.04 in undisturbed plots but there was no significant difference between disturbed and undisturbed plots according to an ANOVA ($F(1,33) = 1.54$, $p = 0.22$). 11 species were found only in the disturbed plots and not in the undisturbed plots. The most common of these were *Combretum celastroides* ($n = 30$), *Acacia reficiens* ($n = 14$), and *Gardenia ternifolia* ($n = 11$). 7 were found only in undisturbed plots, the most common being *Brachystegia spiciformis* ($n = 61$), *Baikiaea plurijuga* ($n = 43$) and *Combretum apiculatum* ($n = 9$). Mean basal area was higher in undisturbed plots (0.5 ± 0.07 m^2 ha^{-1}) than disturbed plots (0.5 ± 0.1 m^2 ha^{-1}).

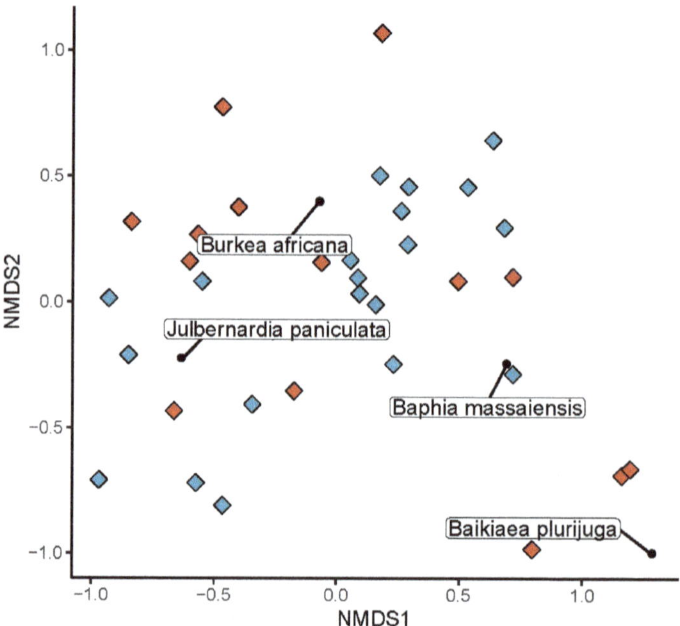

Figure 6. NMDS ordination of species composition of 20×50 m (0.1 ha) plots showing plot scores as coloured diamonds located in disturbed (blue) and undisturbed (red) areas of woodland in Bicuar National Park.

Mean stem density was higher in disturbed plots (900 ± 338.36 stems ha^{-1}) than undisturbed plots (520.3 ± 220.22 stems ha^{-1}). The stem diameter abundance distribution in disturbed plots showed that many more stems were from the 5–10 cm diameter class in disturbed plots, while the disturbed plots had fewer stems in the 10–20 cm size class. Both disturbed and undisturbed plots had a similar abundance of stems in larger stem diameter classes (Figure 8). Multi-stemmed trees in disturbed plots tended to have a greater number of stems per tree (3.4 ± 2.35) than multi-stemmed trees in undisturbed plots (2.4 ± 0.8).

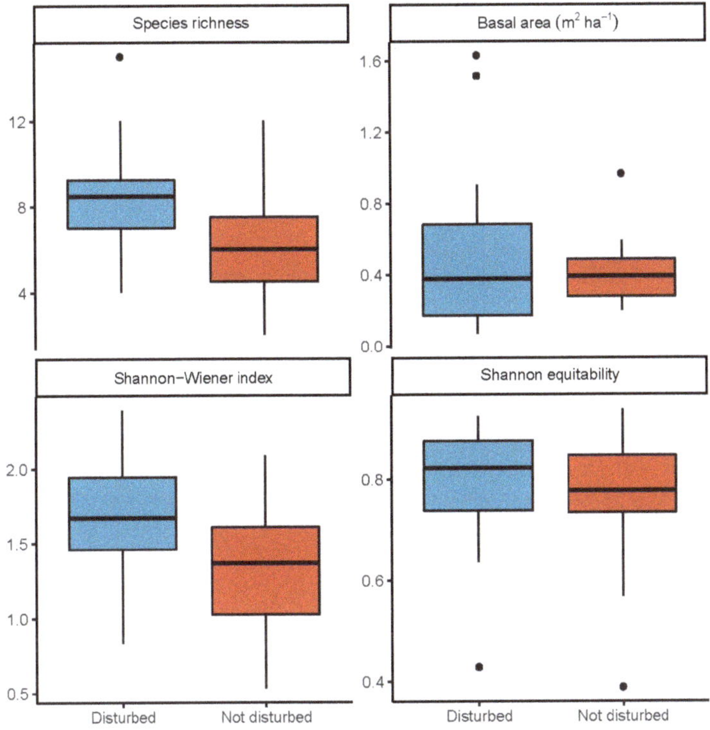

Figure 7. The variation in diversity and woodland structure between disturbed and undisturbed 20 × 50 m (0.1 ha) plots in Bicuar National Park. Boxes bound the 1st and 3rd quartiles, with the median within the box. Whiskers represent 1.5 times the interquartile range plus or minus the 1st and 3rd quartiles, respectively. Values found beyond the whiskers are shown individually as points.

Table 5. Results of ANOVA tests for alpha diversity metrics and plot basal area, between disturbed and undisturbed plots in Bicuar National Park. Mean values for each group of plots with standard errors in parentheses are shown. Asterisks indicate the *p*-value of individual sites in each ANOVA (*** <0.001, ** <0.01, * <0.05, . <0.1).

	Dependent Variable:			
	Species Richness	**Basal Area**	**Shannon (H')**	**Shannon Equit. (E_H)**
Disturbed	2.450 ***	0.098	0.372 **	0.035
	(0.859)	(0.122)	(0.140)	(0.045)
Constant	6.200 ***	0.416 ***	1.311 ***	0.756 ***
	(0.650)	(0.092)	(0.106)	(0.034)
Observations	35	35	35	35
R^2	0.198	0.019	0.176	0.018
Residual Std. Error (df = 33)	2.516	0.357	0.410	0.131
F Statistic (df = 1; 33)	8.126 ***	0.639	7.040 **	0.617

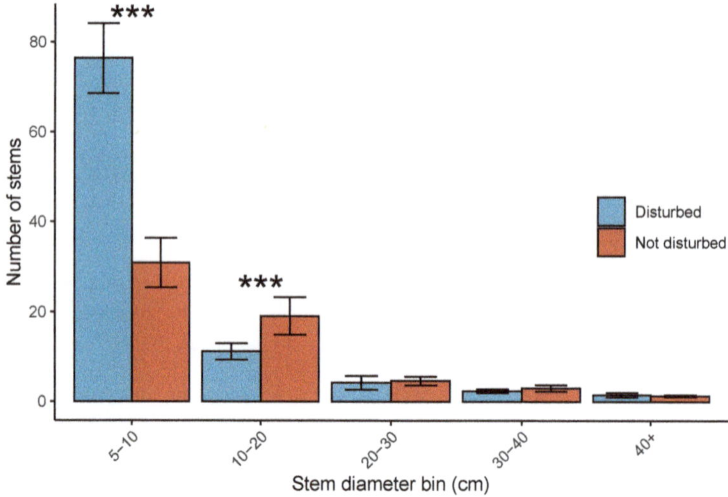

Figure 8. Ranked variation between disturbed and undisturbed plots in stem number, with bars according to stem diameter class. Error bars are the mean ± 1 standard error. Asterisks above pairs of bars refer to the *p*-values of Poisson general linear models which tested whether disturbed and undisturbed plots differ in the number of stems for different stem diameter classes (*** <0.001, ** <0.01, * <0.05, . <0.1).

4. Discussion

4.1. Comparison of Bicuar National Park with Other Woodlands across the Miombo Ecoregion

We compared the tree species diversity and woodland structure of arid woodlands in Bicuar National Park in southwest Angola with three other woodland sites across the miombo ecoregion. Our results show that Bicuar National Park is distinct in both woodland structure and species composition from these other woodlands. Notably, plots in Bicuar National Park contained 27 tree species which did not occur at other sites. This lends support for the Huíla Plateau as an important area for conservation of southern African woodland landscapes. The woodlands in Bicuar National Park were of low tree basal area, with few large trees except in plots dominated by *Baikiaea plurijuga*. Many other studies have drawn a relationship between water availability and basal area [46,47], and our study supports this, with Bicuar National Park being the most arid of the four sites considered in our study. The NMDS of species composition also suggests that plots in Bicuar National Park are influenced by aridity. While there are more arid woodlands within southern Africa, with Mopane woodlands for example often being particularly dry, these plots in Bicuar National park represent particularly dry miombo woodlands.

4.2. Delineation of Woodland Types within Bicuar National Park

Within Bicuar National Park, three distinct woodland types were identified. The first, dominated by *Baikiaea plurijuga* and *Baphia massaiensis* represents the Baikiaea woodland type commonly found to the south of the miombo ecoregion [48]. This is supported by Chisingui et al. [23] who also found Baikiaea woodlands as a distinct woodland type in the Park. *B. plurijuga* has been identified as an important species for conservation, being attractive for selective logging due to its large stature [49,50]. The woodlands created by *B. plurijuga* are also an important habitat for elephants (*Loxodonta africana*) [51,52], with Bicuar National Park and Mupa National Park being key refugia for this animal in the Huíla plateau region. The second woodland type, dominated by *Brachystegia tamarindoides* and *Ochna pulchra* represents a form of small stature woodland with a shrubby understorey and sparse canopy trees, which commonly occurs as a result of repeated disturbance by fire, or poor soil structure [53]. The remaining plots resemble the more archetypical miombo woodland with *Julbernardia paniculata*, though with several species not seen in plots further to the east in the miombo ecoregion such as *Strychnos pungens*. This mosaic of woodland types makes Bicuar National Park a valuable reservoir of diversity and strengthens the case for the Park being a key conservation asset within the Huíla plateau and the larger southern African region. While there are regional boundaries between Baikiaea and miombo woodlands [1], within Bicuar National Park it is likely that the mosaic of woodland types has been created by a combination of soil water capacity and disturbance history. Bicuar has a distinct landscape of wide shallow grassy valleys surrounded by woodland on higher ground (Figure 2). On some of these high points the soil is particularly sandy, resembling the Kalahari sand soils found further east and south [20], and these areas coincide with the presence of Baikiaea woodlands [5]. High levels of disturbance by fire in these Baikiaea patches may additionally prevent a transition to an alternative woodland type via the control of sapling growth.

4.3. Comparison of Disturbed and Undisturbed Woodland Plots

Previously disturbed woodlands around the edge of Bicuar National Park were found to share many species with undisturbed plots in the Park, but with some additional species which did not occur in the undisturbed plots. They also lacked notable archetypical miombo species which tend to form larger canopy trees such as *Brachystegia spiciformis* and contained very few *Julbernardia paniculata*, leading to a distinct woodland composition. The species diversity of these disturbed patches was higher on average than was found in the undisturbed plots, a result which has been corroborated by other studies in miombo woodlands [54–56]. Other studies have shown a peak in species richness during woodland regrowth as pioneer species take advantage of a low competition environment, while some later stage woodland species remain as residuals that survived the original disturbance [30,57]. Gonçalves et al. [30] particularly, notes the dominance of *Pericopsis angolensis* and *Combretum* spp. as light-demanding pioneer species, which were found to be abundant in the disturbed plots here. This suggests that reclamation of previously farmed and abandoned land for landscape conservation in this ecological context is a valuable management strategy.

In disturbed plots near the edge of the Park, there was a lack of species which tend to grow to large canopy trees, possibly due to them being repeatedly felled for timber prior to reclamation by the Park, or due to them being unable to recruit into a more open, shrubby woodland. Despite this lack of canopy forming tree species, some disturbed plots had a greater basal area than undisturbed plots, possibly due to high levels of coppicing in these plots or a divergent fire history. Indeed, mean stem density was higher in undisturbed plots. This can lead to species that would otherwise remain small producing a much larger basal area as they grow multiple stems under high disturbance conditions [58]. The most common species in the disturbed plots were *Combretum psidioides*, *Combretum collinum* and *Terminalia sericea*, members of the Combretaceae family, all of which more commonly remain as smaller multi-stemmed trees in disturbed woodlands, rather than growing to larger canopy trees [59]. This result could be considered at odds with other studies which report lower woody biomass in plots that have experienced harvesting (e.g., Muvengwi et al. [60]). It is important to consider however that our study took place in plots that were measured after farming had been abandoned for at least 7 years, with time for regeneration to occur. It is possible that over time tree basal area will decrease as coppiced shrubby trees are replaced by core miombo species in the transition back to miombo woodland [30]. Indeed, other studies in miombo woodlands across the ecoregion have reported substantial recovery within seven years, with high levels of biomass accumulation in previously disturbed plots [30,61]. Bicuar National Park offers a valuable case study to track woodland regeneration in real time over the next decade in these previously farmed and now protected woodland plots, which could improve our understanding of this potential post-disturbance peak in basal area.

In conclusion, the woodlands of Bicuar National Park represent an important woodland refuge at the far western extent of the miombo ecoregion. These woodlands, both those disturbed by previous farming activity and those which remain undisturbed, possess several species not found commonly in other miombo woodland plots around the region. They may also house important genetic variation for widespread species, representing populations adapted to more arid conditions. Our study highlights the variation in species composition across the miombo ecoregion and underlines the need for studies which incorporate plot data from multiple locations to reach generalisable conclusions about the region as a whole. Additionally, the installation of 15 one hectare woodland monitoring plots and a further twenty 20 × 50 m plots in previously farmed and now protected land offer a valuable natural laboratory to further explore the dynamics of dry miombo woodlands of the Huíla plateau. Bicuar National Park should be considered a key conservation asset within the Huíla plateau and within the miombo ecoregion as a whole, as a successfully protected example of an arid woodland mosaic.

Author Contributions: Investigation and project administration was conducted by J.L.G., F.M.G., J.J.T. and A.V.C. (Bicuar National Park), C.M.R. (Tanzania, Mozambique), J.I.M. and M.N.S. (DRC). The study was conceived by J.L.G. and K.G.D. Data curation, methodology, formal analysis and writing–original draft preparation was conducted by J.L.G. All authors contributed to writing–review and editing. All authors have read and agreed to the published version of the manuscript.

Funding: Final data preparation across all sites was funded by SEOSAW (a Socio-Ecological Observatory for the Southern African Woodlands), a NERC-funded project (Grant No. NE/P008755/1). The installation of woodland plots in Bicuar National Park and their data collection was funded by the National Geographic Society (Grant No. EC-51464R-18) to FMG, AVC, KGD and JLG. JLG was supported by a NERC E3 Doctoral Training Programme PhD studentship (Grant No. NE/L002558/1). The APC was funded by the University of Edinburgh.

Acknowledgments: The rangers at Bicuar National Park are gratefully acknowledged for their help in installing the woodland survey plots and for their help with numerous other incidental challenges during fieldwork. Domingos Fortunato P. Félix da Silva, Abel C. E. Cahali, Felisberto Gomes Armando, José Camôngua Luís, Manuel Jundo Cachissapa and Henrique Jacinto are acknowledged for their help in conducting plot measurements in Bicuar National Park.

Conflicts of Interest: The authors declare no conflict of interest. The funders had no role in the design of the study; in the collection, analyses, or interpretation of data; in the writing of the manuscript, or in the decision to publish the results.

Abbreviations

The following abbreviations are used in this manuscript:

ANOVA	Analysis of Variance
DD	Decimal Degrees
MAP	Mean Annual Precipitation
MAT	Mean Annual Temperature
MAT SD	Standard Deviation of Mean Annual Temperature (Seasonality)
NMDS	Non-metric Multidimensional Scaling
NP	National Park

Appendix A. Estimation of Stem Diameter at 1.3 m via Tree Taper

```r
##' @author Casey M. Ryan
##' @return d130, the~estimated diameter at a POM of 1.3 m (in cm).
##' @param d_in the diameter measured at the POM (in cm)
##' @param POM the height of the POM (in m)
##' @details The adjustment based on tree taper model developed as part of
##'    the ACES project (Abrupt Changes in Ecosystem Services
##'    https://miomboaces.wordpress.com/), using data from the miombo of Niassa.
##'    The model is a cubic polynomial, with~three equations for different sized stems
##' @section Warning: POMs >1.7 m are not adjusted.
POMadj <- function(d_in, POM) {
  stopifnot(is.numeric(d_in),
    is.numeric(POM),
    POM >= 0,
    sum(is.na(POM))==0,
    length(POM) == length(d_in))
  if (any(POM > 1.7))
    warning("POMs >1.7 m are outside the calibration data, no correction applied")
  NAS <- is.na(d_in)
  d_in_clean <- d_in[!NAS]
  POM_clean <- POM[!NAS]
  # define the size class edges:
  edges <- c(5.0, 15.8, 26.6, 37.4)
  sm <- d_in_clean < edges[2]
  med <- d_in_clean >= edges[2] & d_in_clean < edges[3]
  lg <- d_in_clean >= edges[3]

  # compute predictions for delta_d, for~all size classes
  delta_d <- data.frame(
    # if small:
    small =  3.4678+-5.2428 *
      POM_clean + 2.9401 *
      POM_clean^2+-0.7141 *
      POM_clean^3,
    # if med
    med =  4.918+-8.819 *
      POM_clean + 6.367 *
      POM_clean^2+-1.871 *
      POM_clean^3,
    # if large
    large =  9.474+-19.257 *
      POM_clean + 12.873 *
      POM_clean^2+-3.325 *
      POM_clean^3
  )
  # index into the right size class
  dd <- NA_real_
  dd[sm] <- delta_d$small[sm]
  dd[med] <- delta_d$med[med]
  dd[lg] <- delta_d$large[lg]
  dd[POM_clean > 1.7] <- 0  # to avoid extrapolation~mess

  # add NAs back in
  d130 <- NA
  d130[NAS] <- NA
  d130[!NAS] <- d_in_clean -~dd

  if (any(d130[!NAS] < 0))
    warning("Negative d130 estimated, replaced with NA")
  d130[d130 <= 0 & !is.na(d130)] <- NA
  return(d130)
}
```

References

1. White, F. *The Vegetation of Africa: A Descriptive Memoir to Accompany the UNESCO/AETFAT/UNSO Vegetation Map of Africa*; The United Nations Educational, Scientific and Cultural Organization (UNESCO): Paris, France, 1983. [CrossRef]
2. Mayaux, P.; Bartholomé, E.; Fritz, S.; Belward, A. A new land-cover map of Africa for the year 2000. *J. Biogeogr.* **2004**, *31*, 861–877. [CrossRef]
3. Arino, O.; Perez, J.R.; Kalogirou, V.; Defourny, P.; Achard, F. Globcover 2009. In Proceedings of the ESA Living Planet Symposium, Bergen, Norway, 28 June–2 July 2010; pp. 1–3.
4. Chidumayo, E. *Miombo Ecology and Management: An Introduction*; Intermediate Technology Publications: London, UK, 1997.
5. Campbell, B.M.; Jeffrey, S.; Kozanayi, W.; Luckert, M.; Mutamba, M. *Household Livelihoods in Semi-Arid Regions: Options and Constraints*; Center for International Forestry Research: Bogor, Indonesia, 2002; p. 153.
6. Azani, N.; Babineau, M.; Bailey, C.D.; Banks, H.; Barbosa, A.R.; Pinto, R.B.; Boatwright, J.S.; Borges, L.M.; Brown, G.K.; Bruneau, A.; et al. A new subfamily classification of the Leguminosae based on a taxonomically comprehensive phylogeny: The Legume Phylogeny Working Group (LPWG). *Taxon* **2017**, *66*, 44–77. [CrossRef]
7. Privette, J.L.; Tian, Y.; Roberts, G.; Scholes, R.J.; Wang, Y.; Caylor, K.K.; Frost, P.; Mukelabai, M. Vegetation structure characteristics and relationships of Kalahari woodlands and savannas. *Glob. Chang. Biol.* **2004**, *10*, 281–291. [CrossRef]
8. Caylor, K.K.; Dowty, P.R.; Shugart, H.H.; Ringrose, S. Relationship between small-scale structural variability and simulated vegetation productivity across a regional moisture gradient in southern Africa. *Glob. Chang. Biol.* **2004**, *10*, 374–382. [CrossRef]
9. Chidumayo, E.N. Changes in miombo woodland structure under different land tenure and use systems in central Zambia. *J. Biogeogr.* **2002**, *29*, 1619–1626. [CrossRef]
10. Ratnam, J.; Bond, W.J.; Fensham, R.J.; Hoffmann, W.A.; Archibald, S.; Lehmann, C.E.R.; Anderson, M.T.; Higgins, S.I.; Sankaran, M. When is a 'forest' a savanna, and why does it matter? *Glob. Ecol. Biogeogr.* **2011**, *20*, 653–660. [CrossRef]
11. Dexter, K.G.; Smart, B.; Baldauf, C.; Baker, T.R.; Balinga, M.P.B.; Brienen, R.J.W.; Fauset, S.; Feldpausch, T.R.; Ferreira-da Silva, L.; Muledi, J.I.; et al. Floristics and biogeography of vegetation in seasonally dry tropical regions. *Int. For. Rev.* **2015**, *17*, 10–32. [CrossRef]
12. Oliveras, I.; Malhi, Y. Many shades of green: The dynamic tropical forest-savannah transition zones. *Philos. Trans. R. Soc. B Biol. Sci.* **2016**, *371*, 1–15. [CrossRef]
13. Dantas, V.L.; Hirota, M.; Oliveira, R.S.; Pausas, J.G. Disturbance maintains alternative biome states. *Ecol. Lett.* **2016**, *19*, 12–19. [CrossRef]
14. Banda-R, K.; Delgado-Salinas, A.; Dexter, K.G.; Linares-Palomino, R.; Oliveira-Filho, A.; Prado, D.; Pullan, M.; Quintana, C.; Riina, R.; Rodríguez, G.M.; et al. Plant diversity patterns in neotropical dry forests and their conservation implications. *Science* **2016**, *353*, 1383–1387. [CrossRef]
15. Torello-Raventos, M.; Feldpausch, T.R.; Veenendaal, E.; Schrodt, F.; Saiz, G.; Domingues, T.F.; Djagbletey, G.; Ford, A.; Kemp, J.; Marimon, B.S.; et al. On the delineation of tropical vegetation types with an emphasis on forest/savanna transitions. *Plant Ecol. Divers.* **2013**, *6*, 101–137. [CrossRef]
16. Frost, P. The ecology of miombo woodlands. In *The Miombo in Transition: Woodlands and Welfare in Africa*; Campbell, B., Ed.; Center for International Forestry Research: Bogor, Indonesia, 1996; pp. 11–55.
17. Ryan, C.M.; Pritchard, R.; McNicol, I.; Owen, M.; Fisher, J.A.; Lehmann, C. Ecosystem services from southern African woodlands and their future under global change. *Philos. Trans. R. Soc. B Biol. Sci.* **2016**, *371*, 1–16. [CrossRef] [PubMed]
18. Mayaux, P.; Eva, H.; Brink, A.; Achard, F.; Belward, A. Remote sensing of land-cover and land-use dynamics. In *Earth Observation of Global Change: The Role of Satellite Remote Sensing in Monitoring the Global Environment*; Springer: Berlin, Germany, 2008; pp. 85–108._5. [CrossRef]
19. Clarke, D.A.; York, P.H.; Rasheed, M.A.; Northfield, T.D. Does biodiversity-ecosystem function literature neglect tropical ecosystems. *Trends Ecol. Evol.* **2017**, *32*, 320–323. [CrossRef] [PubMed]
20. Huntley, B.J.; Lages, F.; Russo, V.; Ferrand, N. (Eds.) *Biodiversity of Angola: Science & Conservation: A Modern Synthesis*; Springer: Cham, Switzerland, 2019. [CrossRef]

21. Soares de Oliveira, R. *Magnificent and Beggar Land: Angola since the Civil War*; Hurst Publishers: London, UK, 2015.
22. Figueiredo, E.; Smith, G.F.; César, J. The flora of Angola: First record of diversity and endemism. *Taxon* **2009**, *58*, 233–236. [CrossRef]
23. Chisingui, A.V.; Gonçalves, F.M.P.; Tchamba, J.J.; Camôngua, L.J.; Rafael, M.F.F.; Alexandre, J.L.M. *Vegetation Survey of the Woodlands of Huíla Province*; Klaus Hess Publishers: Gottingen, Germany; Windhoek, Namibia, 2018. [CrossRef]
24. Soares, M.; Abreu, J.; Nunes, H.; Silveira, P.; Schrire, B.; Figueiredo, E. The leguminosae of Angola: Diversity and endemism. *Syst. Geogr. Plants* **2007**, *77*, 141–212. [CrossRef]
25. Figueiredo, E.; Smith, G.F. Plants of Angola/Plantas de Angola. *Strelitzia* **2008**, *22*, 1–279.
26. Linder, H.P. Plant diversity and endemism in sub-Saharan tropical Africa. *J. Biogeogr.* **2001**, *28*, 169–182. [CrossRef]
27. Droissart, V.; Dauby, G.; Hardy, O.J.; Deblauwe, V.; Harris, D.J.; Janssens, S.; Mackinder, B.A.; Blach-Overgaard, A.; Sonké, B.; Sosef, M.S.M.; et al. Beyond trees: Biogeographical regionalization of tropical Africa. *J. Biogeogr.* **2018**, *45*, 1153–1167. [CrossRef]
28. Schneibel, A.; Stellmes, M.; Revermann, R.; Finckh, M.; Röder, A.; Hill, J. Agricultural expansion during the post-civil war period in southern Angola based on bi-temportal Landsat data. *Biodivers. Ecol.* **2013**, *5*, 311–319. [CrossRef]
29. McNicol, I.M.; Ryan, C.M.; Williams, M. How resilient are African woodlands to disturbance from shifting cultivation? *Ecol. Appl.* **2015**, *25*, 2320–2336. [CrossRef]
30. Gonçalves, F.M.P.; Revermann, R.; Gomes, A.L.; Aidar, M.P.M.; Finckh, M.; Juergens, N. Tree species diversity and composition of Miombo woodlands in South-Central Angola: A chronosequence of forest recovery after shifting cultivation. *Int. J. For. Res.* **2017**, *2017*, 1–13. [CrossRef]
31. Ryan, C.M.; Williams, M.; Grace, J. Above- and belowground carbon stocks in a miombo woodland landscape of Mozambique. *Biotropica* **2011**, *43*, 423–432. [CrossRef]
32. McNicol, I.M.; Ryan, C.M.; Dexter, K.G.; Ball, S.M.J.; Williams, M. Aboveground carbon storage and its links to stand structure, tree diversity and floristic composition in south-eastern Tanzania. *Ecosystems* **2018**, *21*, 740–754. [CrossRef] [PubMed]
33. Muledi, J.I.; Bauman, D.; Drouet, T.; Vleminckx, J.; Jacobs, A.; Lejoly, J.; Meerts, P.; Shutcha, M.N. Fine-scale habitats influence tree species assemblage in a miombo forest. *J. Plant Ecol.* **2017**, *10*, 958–969. [CrossRef]
34. Fick, S.E.; Hijmans, R.J. WorldClim 2: New 1-km spatial resolution climate surfaces for global land areas. *Int. J. Climatol.* **2017**, *37*, 4302–4315. [CrossRef]
35. Conservatoire et Jardin Botaniques de la Ville de Genève and South African National Biodiversity Institute. African Plant Database (Version 3.4.0). 2020. Available online: https://www.ville-ge.ch/musinfo/bd/cjb/africa/recherche.php (accessed on 5 November 2019).
36. Palgrave, K.C. *Trees of Southern Africa*; Struik Publications: Cape Town, South Africa, 2003.
37. Kershaw, J.A.; Ducey, M.J.; Beers, T.W.; Husch, B. *Forest Mensuration*; John Wiley & Sons: Chichester, UK, 2017.
38. UNEP-WCMC and IUCN. *Protected Planet: The World Database on Protected Areas (WDPA)*. 2019. Available online: https://www.protectedplanet.net/ (accessed on 15 January 2019).
39. Kahle, D.; Wickham, H. ggmap: Spatial visualization with ggplot2. *R J.* **2013**, *5*, 144–161. [CrossRef]
40. Chave, J.; Réjou-Méchain, M.; Búrquez, A.; Chidumayo, E.; Colgan, M.S.; Delitti, W.B.C.; Duque, A.; Eid, T.; Fearnside, P.M.; Goodman, R.C.; et al. Improved allometric models to estimate the aboveground biomass of tropical trees. *Glob. Chang. Biol.* **2014**, *20*, 3177–3190. [CrossRef]
41. Oksanen, J.; Blanchet, F.G.; Friendly, M.; Kindt, R.; Legendre, P.; McGlinn, D.; Minchin, P.R.; O'Hara, R.B.; Simpson, G.L.; Solymos, P.; et al. *vegan: Community Ecology Package*; R Package Version 2.5-5; 2019. Available online: https://CRAN.R-project.org/package=vegan/ (accessed on 15 January 2019).
42. Koleff, P.; Gaston, K.J.; Lennon, J.J. Measuring beta diversity for presence-absence data. *J. Anim. Ecol.* **2003**, *72*, 367–382. [CrossRef]
43. Smith, B.; Wilson, J.B. A consumer's guide to evenness indices. *Oikos* **1996**, *76*, 70–82. [CrossRef]
44. Legendre, P.; De Cáceres, M. Beta diversity as the variance of community data: Dissimilarity coefficients and partitioning. *Ecol. Lett.* **2013**, *16*, 951–963. [CrossRef]
45. R Core Team. *R: A Language and Environment for Statistical Computing*; R Foundation for Statistical Computing: Vienna, Austria, 2019.

46. Terra, M.C.N.S.; Santos, R.M.D.; Prado Júnior, J.A.; de Mello, J.M.; Scolforo, J.R.S.; Fontes, M.A.L.; ter Steege, H. Water availability drives gradients of tree diversity, structure and functional traits in the Atlantic-Cerrado-Caatinga transition, Brazil. *J. Plant Ecol.* **2018**, *11*, 803–814. [CrossRef]

47. Strickland, C.; Liedloff, A.C.; Cook, G.D.; Dangelmayr, G.; Shipman, P.D. The role of water and fire in driving tree dynamics in Australian savannas. *J. Ecol.* **2016**, *104*, 828–840. [CrossRef]

48. Timberlake, J.; Chidumayo, E.; Sawadogo, L. Distribution and characteristics of African dry forests and woodlands. In *The Dry Forests and Woodlands of Africa: Managing for Products and Services*; EarthScan: London, UK, 2010; pp. 11–42.

49. Ng'andwe, P.; Chungu, D.; Shakacite, O.; Vesa, L. Abundance and distribution of top five most valuable hardwood timber species in Zambia and their implications on sustainable supply. In Proceedings of the 6th International Conference on Hardwood Processing, Lahti, Finland, 25–28 September 2017; Mottonen, V., Heinonen, E., Eds.; Natural Resources Institute Finland: Helsinki, Finland, 2017; pp. 18–27.

50. Wallenfang, J.; Finckh, M.; Oldeland, J.; Revermann, R. Impact of shifting cultivation on dense tropical woodlands in southeast Angola. *Trop. Conserv. Sci.* **2015**, *8*, 863–892. [CrossRef]

51. Sianga, K.; Fynn, R. The vegetation and wildlife habitats of the Savuti-Mababe-Linyati ecosystem, northern Botswana. *KOEDOE* **2017**, *59*, 1–16. [CrossRef]

52. Mukwashi, K.; Gandiwa, E.; Kativu, S. Impact of African elephants on *Baikiaea Plurijuga* woodland around natural and artificial watering points in northern Hwange National Park, Zimbabwe. *Int. J. Environ. Sci.* **2012**, *2*, 1355–1368. [CrossRef]

53. Smith, P.; Allen, Q. *Field Guide to the Trees and Shrubs of the Miombo Woodlands*; Royal Botanic Gardens, Kew: London, UK, 2004.

54. Caro, T.M. Species richness and abundance of small mammals inside and outside and African national park. *Biol. Conserv.* **2001**, *98*, 251–257. [CrossRef]

55. McNicol, I.M.; Ryan, C.M.; Mitchard, E.T.A. Carbon losses from deforestation and widespread degradation offset by extensive growth in African woodlands. *Nat. Commun.* **2018**, *9*, 1–11. [CrossRef]

56. Shackleton, C.M. Comparison of plant diversity in protected and communal lands in the Bushbuckridge lowveld savanna, South Africa. *Biol. Conserv.* **2000**, *94*, 273–285. [CrossRef]

57. Kalaba, F.K.; Quinn, C.H.; Dougill, A.J.; Vinya, R. Floristic composition, species diversity and carbon storage in charcoal and agriculture fallows and management implications in Miombo woodlands of Zambia. *For. Ecol. Manag.* **2013**, *304*, 99–109. [CrossRef]

58. Luoga, E.J.; Witkowski, E.T.F.; Balkwill, K. Regeneration by coppicing (resprouting) of miombo (African savanna) trees in relation to land use. *For. Ecol. Manag.* **2004**, *189*, 23–25. [CrossRef]

59. van Wyk, B.; van Wyk, P. *Field Guide to Trees of Southern Africa*; Struik Nature: Cape Town, South Africa, 2014.

60. Muvengwi, J.; Chisango, T.; Mpakairi, K.; Mbiba, M.; Witkowski, E.T.F. Structure, composition and regeneration of miombo woodlands within harvested and unharvested areas. *For. Ecol. Manag.* **2020**, *458*, 1–10. [CrossRef]

61. Chidumayo, E.N. Forest degradation and recovery in a miombo woodland landscape in Zambia: 22 years of observations on permanent sample plots. *For. Ecol. Manag.* **2013**, *291*, 154–161. [CrossRef]

Article

Diversification of African Rainforest Restricted Clades: Piptostigmateae and Annickieae (Annonaceae)

Baptiste Brée [1], Andrew J. Helmstetter [1], Kévin Bethune [1], Jean-Paul Ghogue [2], Bonaventure Sonké [2] and Thomas L. P. Couvreur [1,*]

[1]   IRD, DIADE, CIRAD, University of Montpellier, 34090 Montpellier, France; baptiste.bree@etu.umontpellier.fr (B.B.); andrew.j.helmstetter@gmail.com (A.J.H.); kevin.bethune@hotmail.fr (K.B.)

[2]   Plant Systematic and Ecology Laboratory, Department of Biology, Higher Teachers' Training College, University of Yaoundé I, P.O. Box 047, Yaoundé, Cameroon; jpghogue@greenconnexion-cm.org (J.-P.G.); bonaventuresonke@ens.cm (B.S.)

*   Correspondence: Thomas.Couvreur@ird.fr

Received: 15 May 2020; Accepted: 5 June 2020; Published: 7 June 2020

Abstract: African rainforests (ARFs) are species rich and occur in two main rainforest blocks: West/Central and East Africa. This diversity is suggested to be the result of recent diversification, high extinction rates and multiple vicariance events between west/central and East African forests. We reconstructed the diversification history of two subtribes (Annickieae and Piptostigmateae) from the ecologically dominant and diverse tropical rainforest plant family Annonaceae. Both tribes contain endemic taxa in the rainforests of West/Central and East Africa. Using a dated molecular phylogeny based on 32 nuclear markers, we estimated the timing of the origin of East African species. We then undertook several diversification analyses focusing on Piptostigmateae to infer variation in speciation and extinction rates, and test the impact of extinction events. Speciation in both tribes dated to the Pliocene and Pleistocene. In particular, *Piptostigma* (13 species) diversified mainly during the Pleistocene, representing one of the few examples of Pleistocene speciation in an African tree genus. Our results also provide evidence of an ARF fragmentation at the mid-Miocene linked to climatic changes across the region. Overall, our results suggest that continental-wide forest fragmentation during the Neogene (23.03–2.58 Myr), and potentially during the Pliocene, led to one or possibly two vicariance events within the ARF clade Piptostigmateae, in line with other studies. Among those tested, the best fitting diversification model was the one with an exponential speciation rate and no extinction. We did not detect any evidence of mass extinction events. This study gives weight to the idea that the ARF might not have been so negatively impacted by extinction during the Neogene, and that speciation mainly took place during the Pliocene and Pleistocene.

Keywords: biogeographic vicariance; extinction; phylogenomics; gene shopping; gene capture; molecular dating

---

1. Introduction

Tropical rainforests are the most biodiverse terrestrial ecosystems in the world, despite covering less than 10% of the land surface [1,2]. African rainforests (ARFs) are one of the world's most biodiverse regions [3–5] and the Congo basin contain the second largest continuous expanse of this biome after the Amazon basin. Understanding why and when the ARF flora diversified has been the subject of several studies [6], which highlighted two main geological periods that could have been associated with increased speciation rates in ARF clades: (1) Late Oligocene–Miocene, characterized by climatic fluctuations and geological events that caused the fragmentation and re-expansion of ARFs, leading to increased allopatric speciation [7–10]; (2) the Pleistocene, with paleobotanical studies suggesting that rapid climatic changes linked to the Milankovitch cycles enhanced allopatric speciation by isolating ARFs into small patches (lowland forest refugia) [11–14].

In addition, even though ARFs are species rich, they are less diverse than other major tropical rainforested regions such as the Neotropics and South East Asia, a pattern of tropical biodiversity referred to as the "odd man out" [15]. Numerous hypotheses have been suggested to explain these differences (reviewed by [16]). One hypothesis is linked to differences in diversification rates between regions, with tropical Africa undergoing higher levels of extinction rates connected to increased aridification since the Miocene [2,16–18]. Inferring diversification variation within rainforest restricted plant clades spanning most of the Cenozoic can provide a way to unravel the importance of the Miocene and Pleistocene periods, and also help understand the possible impact of past extinction history on present ARF diversity.

In parallel to the above, dated molecular phylogenies have also provided insights into ARF cover history since the Oligocene. Today, ARFs are divided into two main disjunct regions separated by a 1000 km wide arid corridor running from north to south in East Africa [9,19]: the Guineo–Congolian region (GC, located in west and central Africa) and the East African region (EA). In the former, forests extend from Liberia to the Republic of Congo and in vast areas of the Congo Basin. In the latter, rainforests occur in small patches along the East African coast and the Eastern Arc mountains of Kenya and Tanzania, and reaching northern Mozambique (Figure 1). Many floristic affinities have been suggested between these two biogeographic regions: many genera are widespread and common to both regions but at the specific level, most species are endemic either to GC or EA. In addition, each region is home to several endemic genera [9,20–26]. These affinities suggest that rainforests were once widespread across the continent, leading to a pan-African rainforest. Numerous studies suggested that ARFs could have stretched across the width of the continent during Paleocene and Eocene [2,27–29] (but see [30]).

Figure 1. Present day distribution of tropical African rainforests separated between: (in red) the Guineo–Congolian region occurring in West to central Africa, and (in blue) the East African coastal and Eastern Arc rainforests. Map adapted from Couvreur et al. (2008) [9]. Photos represent different species from the Piptostigmateae tribe. A to G: endemic species to the Guineo–Congolian rainforests; (**H, I**) endemics to the east African rainforests. (**A**) *Sirdavidia solannona*, flower; (**B**) *Greenwayodendron suaveolens*, flower; (**C**) *Polyceratocarpus parviflorus*, flower; (**D**) *Annickia affinis*, fruit; (**E**) *Annickia affinis*, flower; (**F**) *Piptostigma multinervium*, flower; (**G**) *Piptostigma pilosum*, leaf, note numerous secondary veins; (**H**) *Mwasumbia alba*, flower; (**I**) *Annickia kummeriae*, flower. Photos: Thomas L. P. Couvreur.

Two hypotheses have been proposed as to how these disjunct patterns emerged: (1) the ARF fragmented only once during the Oligocene or Early Miocene due to global aridification following climatic changes of the Eocene/Oligocene Transition (EOT) around 33 Million years ago (Myr) [17,23,31], (2) ARFs have fragmented repeatedly since the Oligocene with phases of continental expansion and retraction linked to major climatic events [2,22,32,33]. Using a dated molecular phylogeny, Couvreur et al. (2008) [9] showed support for the second hypothesis, as they linked multiple vicariance speciation events with known periods of aridification and geological events in the rainforest restricted clade of African Annonaceae (tribe Monodoreae), revealing three possible fragmentation events during the Miocene and Oligocene epochs (at ca. 33, ca. 16 and ca. 8 Myr). Most other molecular dating studies published to date also support the hypothesis of multiple fragmentations in plant and animal clades, suggesting a shared pattern [34,35]. By focusing on an ARF restricted plant clade of the Annonaceae family, we seek, in this study, to infer how many vicariance events occurred, if these are congruent with known aridification periods and if these events coincide with other independently estimated events in other ARF clades. In this context, studying speciation events through time in endemic ARF clades spanning both separated regions will provide complementary data to test these hypotheses. In order to test these hypotheses and infer vicariance ages, we can use dated molecular phylogenies of clades spanning east and west/central Africa, with endemic species in both regions [9,36].

Annonaceae are a pantropical family of trees, shrubs and lianas that are the most species-rich family of Magnoliales [37,38] and represent an important component of ARF in general [5] and locally [39]. Like numerous clades, Annonaceae are less diverse in Africa compared to the other main tropical rainforests regions, with 1100 species in South East Asia, 930 in the Neotropics and only 400 in Africa. The phylogeny of this family was studied for the first time more than 20 years ago [40], and since then, phylogenetic reconstructions have mainly been based on chloroplast markers [37,38,41,42]. More recently, the first Annonaceae phylogeny using hundreds of nuclear markers from a tailored baiting kit was published by Couvreur et al. (2019) [43], leading to robust phylogenetic inferences at generic and species levels. Such a kit provides a unique opportunity to infer phylogenetic relationships across Annonaceae and enables molecular dating within a single integrated framework.

The tribe Piptostigmateae (Malmeoideae subfamily) consists of six genera (*Brieya*, *Greenwayodendron*, *Mwasumbia*, *Piptostigma*, *Polyceratocarpus* and *Sirdavidia*), containing 31 species [43–49]. This tribe, together with its closely related tribe Annickieae (one genus/eight species [43]), presents an interesting biogeographic pattern, useful for testing the above hypotheses. Three genera contain endemic species in both ARF regions: *Greenwayodendron* (five GC/one EA endemics), *Annickia* (seven GC/one EA endemics) and *Polyceratocarpus* (six GC/two EA endemics). The Piptostigmateae tribe also contains two endemic sister genera with narrow ranges: *Sirdavidia* (GC), from the rainforests of Gabon and Cameroon, and *Mwasumbia* (EA), endemic to the coastal forests of Tanzania [43]. Estimating speciation events within this tribe between east and west/central African endemics and correlating these with known abiotic events responsible for rainforest fragmentation (i.e., aridification events) can improve our understanding of the ARF flora diversification.

Here, we aim to answer three main questions about ARF evolution using the Piptostigmateae and Annickieae tribes: (1) What were the diversification patterns driving Piptostigmateae evolution during the Cenozoic? (2) Has extinction played an important role in shaping current ARF diversity? (3) How many vicariance events do we infer within this clade and are these events congruent with known climatic and geological events?

2. Material and Methods

2.1. Taxon and Nuclear Marker Sampling

We sequenced one individual per species for the 23 available species, out of the 31 species known in Piptostigmateae (74%), covering all six genera of the tribe, and seven species out of eight, for the genus *Annickia* (Annickieae). We made sure to choose species from both ARF regions in the genera

Greenwayodendron and *Annickia* (endemic EA *Polyceratocarpus* species were not available). Our ingroup includes, in total, three endemic species from the East African region, and 27 endemic species from the Guineo–Congolian region (Table S1). Finally, we selected and added 23 species as outgroups from four different subfamilies: eight Malmeoideae (same subfamily as Piptostigmateae), 11 Annonoideae, three Ambaviodeae and one Anaxagoreoideae.

The Annonaceae bait kit [43] was used for DNA sequencing. This kit targets a total of 469 exons Annonaceae wide. Details on all 53 species (Table S2) used in this study, such as field sampling, DNA extraction and DNA sequencing are presented in Couvreur et al. (2019) [43]. We therefore retrieved fastq files from the AFRODYN bioproject (https://www.ncbi.nlm.nih.gov/bioproject/PRJNA508895), which contains demultiplexed, trimmed and sorted DNA paired reads (R1 forward + R2 reverse, 150 bp per read) of more than 120 Annonaceae species. Couvreur et al. (2019) [42] also provided the reference file for the baited regions, a fasta file comprising 469 targeted exons.

2.2. Contig Assembly, Alignment and Paralog Identification

Raw reads from 53 individuals were cleaned as in Couvreur et al. (2019) [43]. We processed our data with HybPiper (v. 1.2) [50] taking, as inputs, the clean fastq reads and the reference file containing the targeted exon sequence data. We retrieved contigs of exons and flanking introns and we aligned them separately with MAFFT (v. 7.305) [51]. Poorly aligned regions were cleaned using Gblocks (v. 0.91b, gaps allowed) [52]. HybPiper flags potential paralogs, which we assessed and removed from downstream analyses as follows. If several contigs covered more than 85% of the length of the same target gene, they were set aside for verification. Phylogenetic trees were created for each group of potential paralogs using RAxML [53]. It might be expected that alternative sequences were grouped by individual (e.g., different alleles in the same individual). However, if sequences corresponding to the "main" contig (i.e., the contig best matching the target reference) and other alternative sequences (numbered from 1 to n) clustered into distinct groups in the tree, these groups were considered as paralogs. In this case, we removed the entire locus from subsequent analyses.

2.3. Exons and Introns Selection for Molecular Dating

To minimize missing data for our phylogenetic inference, we selected exons and introns that were present in more than 75% of individuals and for which we recovered at least 75% of the exon's length (referred to as the 75/75 dataset), following Couvreur et al. (2019) [43]. Firstly, a phylogenetic tree was inferred with RAxML [53] for each of the exons and introns in our 75/75 dataset. These trees were then rooted one by one with the most phylogenetically distant outgroup available. We used TreeShrink [54] with default parameters, to automatically identify and remove outliers in our trees The number and length of loci in our relatively large dataset made it computationally intractable to use all of the data for divergence time estimation. To identify a subset of genetic markers that would provide accurate, computationally feasible estimates of divergence time, we took a "gene-shopping" approach using the SortaDate pipeline [55]. Root-to-tip variances (that is the variance of the distance of branch lengths from root to each tip) were calculated for each tree and used as indicators to determine which trees are most likely to follow a strict molecular clock. This was done to reduce model complexity and select exons and introns with similar (but not identical) evolutionary rates. However, we did not apply the phylogenetic relationship filter, i.e., we did not select genes that produced a particular phylogenetic relationship. A total of 32 sequences (31 exons and 1 introns) with the lowest "root-to-tip" values were selected for the dating analysis. The ModelTest-NG tool [56] was then used to identify the best nucleotide substitution models for each of the 32 sequences, according to the Bayesian Information Criterion (BIC).

2.4. Molecular Dating

The phylogenetic tree was inferred and dated using BEAST (v. 2.5.2) [57], a Bayesian approach for generating time-calibrated phylogenetic trees. Pirie and Doyle (2012) [58] concluded that the oldest reliable age estimate for the Annonaceae crown node is *Futhabanthus*, a fossil flower which dates back to 89 Myr. *Endressinia* provides the most recent common ancestor (MRCA) of Magnoliaceae and Annonaceae at 112 Myr. As there is no other reliable fossil for Annonaceae, and no evidence of Piptostigmateae fossils, we chose to constrain the Annonaceae crown node between those dates (89–112 Myr) and apply a uniform prior following [58]. In addition, we imposed two topological constraints based on previous phylogenetic analyses: *Anaxagorea crassipetala* as the sister to the rest of the family, and *Anaxagorea crassipetala* plus the three sampled species of Ambavioideae as the sisters to the rest of the family [37,43]. After using Nested Sampling [59] to select the best model (see Table S7), we chose an uncorrelated lognormal relaxed molecular clock model, as well as a Yule tree prior model (identical net diversification rate in all branches, no extinction). We ran each analysis for 20 million generations, sampling every 2000 generations and repeated the process three times to compare the consistency of results across runs. We aimed to reach an adequate Effective Sample Size (ESS) of our posterior distribution above 200 for all parameter estimates. The output of each chain was then analyzed using Tracer (v. 1.7.1) [60] to check for convergence between the three runs. LogCombiner (v. 2.5.2) [61] was used to combine converging runs into a single chain (a burnin of 20% for each analysis). TreeAnnotator (v. 2.5.2) [61] was then used to determine the Maximum Clade Credibility (MCC) tree, as well as the mean ages, 95% highest posterior density (HPD) interval and the posterior probability (PP) of each node. We used the results of this dating analysis to find vicariance events across ARF history.

2.5. Diversification Analyses

Diversification analyses were conducted on the Piptostigmateae tribe only, because Annickieae is sister to all Malmeoideae and the sampling within the rest of the tribe (besides Piptostigmateae) is highly incomplete. The 21 outgroups as well as the seven Annickia were thus pruned from the MCC tree using the "drop.tip" function of the package ape (v. 5.3) [62,63]. For each analysis, the number of missing taxa in the phylogeny was specified for each of the following three methods, as our data contained 74% of known Piptostigmatae species. In order to cross validate our results, we undertook three different diversification analyses under maximum likelihood and Bayesian approaches. We first used RPANDA (v. 1.5) [64], to fit different diversification models to the dated phylogenetic tree under a maximum likelihood framework and estimate speciation and extinction rates through time. After several tests to find the best starting parameters (a range of priors between 0.1 and one were tested, with negligible change to the results), we set 0.2 events per lineage per Myr for the speciation rate and 0.05 events per lineage per Myr for the extinction rate. We then chose to test six different diversification models: two null models (time constant birth model and time constant birth–death model), and four time-dependent models: (a) pure-birth (no extinction) exponential speciation rate, (b) birth–death with exponential speciation rate and constant extinction rate, (c) birth–death with constant speciation rate and exponential extinction rate, and (d) birth–death with exponential speciation and extinction rates. The choice of the best model was made using the corrected Akaike information criterion (AICc) following the recommendations of Burnham and Anderson (2002) [65].

Secondly, in order to conduct a comparison with RPANDA results, we used a Bayesian approach as implemented in TESS (v. 2.1.0) [66] to fit the same six diversification models as above and compute their marginal likelihoods. We set an exponential prior distribution with rate = 10 for the speciation and extinction rates for all models, and ran Markov chain Monte Carlo (MCMC) analyses (100,000 iterations, 10,000 burn-in). We then used steppingstone simulations to estimate the marginal likelihood for each of our models (100 stepping-stone iterations, 10 burnin). MCMC convergence was verified with the coda package to ensure that ESS values were at least above 100 (we also verified mixing with a visualization of trace and density of the MCMC), and models were compared using Bayes Factors (BF).

Thirdly, we used BAMM (v ? 5.0) [67,68] to detect possible diversification rate shifts along the branches of the tree. This Bayesian inference method uses a reversible jump Markov chain Monte Carlo (rjMCMC) approach to explore various models, detect speciation rate shifts over time and estimate diversification parameters (i.e., speciation and extinction). Initial parameters were estimated with the R package BAMMtools (v. 2.1.6) [67] (expected number of shifts = 1), and the other priors were set to their default values. We ran a rjMCMC for 10 million generations (sampling every 1000 generations, 10% burnin) in order to ensure the convergence on four Metropolis-coupled MCMC chains (default values). Effective sample sizes for the likelihood and number of shifts were verified (ESS > 200) using the package coda [69]. BF were then calculated with the computeBayesFactor command in BAMMtools to find the best fitting-model.

Finally, to test the hypothesis that the Piptostigmateae tribe was impacted by one or several mass extinctions events during its evolutionary history, we used CoMET implemented in TESS [66]. This Bayesian statistical method computes the joint posterior distribution of speciation-rate shifts, extinction-rate shifts, mass-extinction events and estimate their values by running rjMCMC simulations over multiple episodically varying birth–death models. This method has also been suggested to perform better with small phylogenies, as in our case, when compared to other methods such as BDSKY [70]. We set the priors of mass extinction and speciation rate shifts to one as we had no a priori information. We ran the rjMCMC with a constraint of 500 ESS for each shift minimum, and we considered shifts as significant if 2 ln BF ≥ 6 [71].

3. Results

3.1. Phylogenomics

We recovered all the 469 exons present in the baiting kit with their corresponding introns. Our 75/75 dataset contained 330 exons and 329 introns. A total of 17 exons and 17 introns were flagged by HybPiper as potential paralogs, and of these, 32 were part of the 75/75 dataset. These were discarded, leaving a total of 298 exons and 297 introns for downstream phylogenetic analyses. After using our "gene shopping" approach, the 31 exons and one intron selected based on root-to-tip variance had a total length of 22,001 bp. The appropriate nucleotide substitution models selected for each exon and intron can be found in Table S3. Our three MCMC chains converged to similar values and the combined ESS values was above of 200 for all parameters and greater than 600 for posterior and likelihood values. The majority of nodes were highly supported in the resulting MCC tree (Figure 2), with all PPs over 0.96, and over 80% of nodes having posterior probability (PP) equal to 1.0. Annickieae and Piptostigmateae were recovered as monophyletic with maximum support (PP = 1.0). Finally, our results also support the placement of Annickieae as a sister to the rest of the Malmeoideae subfamily.

3.2. Molecular Dating and Divergence Times

The crown node of Annonaceae was dated at 91.85 Myr (95% HPD interval: 89.00–97.74 Myr). The age of the crown node of the Piptostigmateae tribe is dated to 33.35 Myr (95% HPD interval: 30.00–37.12 Myr). The crown ages of extant lineages for the six Piptostimateae genera and *Annickia* are presented in Table S4. Nodes corresponding to speciation events between endemic species of the EA region (in red in Figure 2) and endemic species of the GC region are dated as follows: divergence between *Annickia kummeriae* and the rest of the Annickieae is estimated at 5.36 Myr (95% HPD interval: 4.63–6.17 Myr). The divergence between *Greenwayodendron usambaricum* and *Greenwayodendron gabonicum, G. glabrum, G. litorale, G. suaveolens* is estimated at 5.79 Myr (95% HPD interval: 4.94–6.68 Myr). Finally, divergence between *Mwasumbia alba* and *Sirdavidia solannona* is estimated at 16.13 Myr (95% HPD interval: 13.89–18.18 Myr). All the nodes described here are well supported with a PP of 1.0.

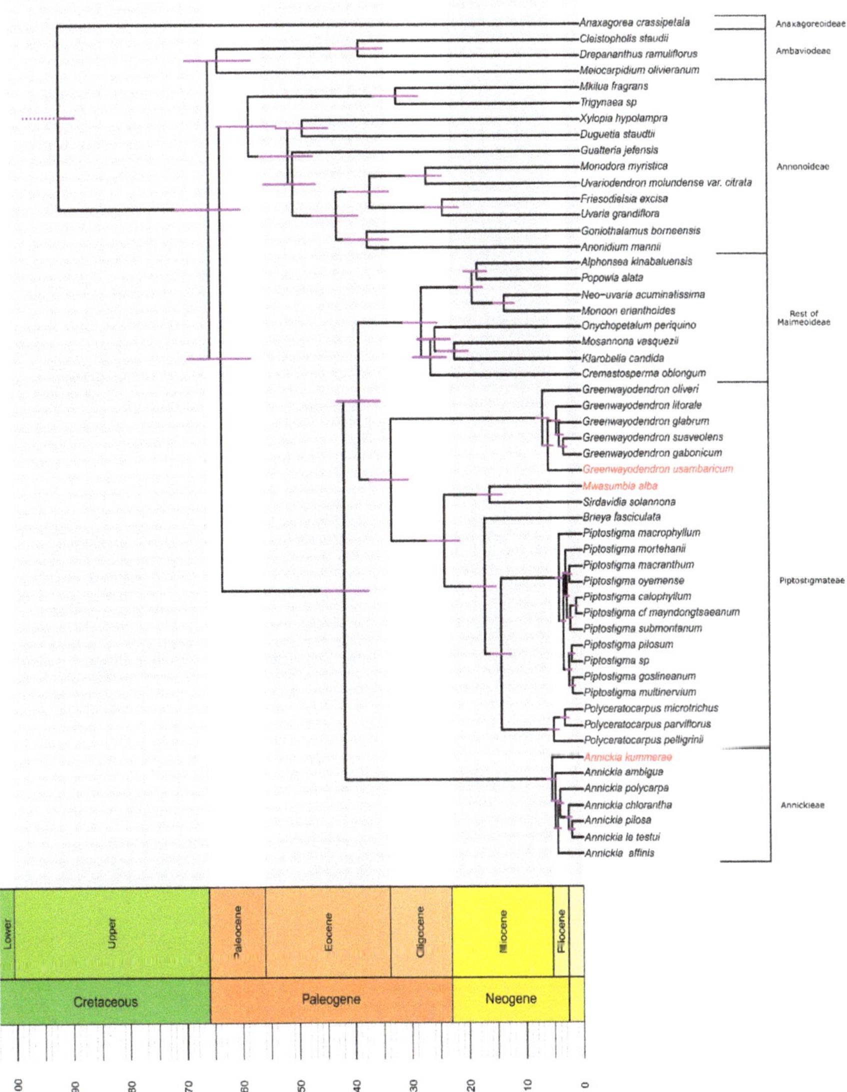

Figure 2. Maximum Clade Credibility (MCC) tree of the tribes Piptostigmateae and the Annickieae, based on 32 nuclear loci. Species colored in red are East African endemics. Purple bars represent the 95% highest posterior density (HPD) interval for each node. The last period corresponds to the Pleistocene epoch.

3.3. Diversification Analyses

Results of the diversification analyses with RPANDA supported the pure-birth (no extinction) exponential speciation rate as the best fitting model for Piptostigmateae, with speciation increasing through time (Table 1). Cross-validation with TESS also supported the pure-birth exponential speciation rate as the best fitting model (Table S5). However, Bayes Factor comparisons did not significantly distinguish this model from other models tested that had constant or no extinction (Table S5, difference in marginal likelihood with the simplest model (constant speciation/no extinction) < 2). Finally, the

BAMM analysis did not detect any significant rate shifts during the evolution of Piptostigmateae. ESS values for the likelihood and the number of shifts were greater than 2900, and BF analysis supported the null model (no shift) as the best fitting model (Table S6). This model indicates that speciation rates have gradually increased through time during the clades history (Figure 3), in agreement with the RPANDA and TESS results (Table 1, Table S5). It also shows a higher (but not significant) speciation rate for the *Piptostigma* genus (mean speciation rate through the clade = 0.2027) compared to others (*Polyceratocarpus* = 0.1845, *Greenwayodendron* = 0.1804, *Mwasumbia* = 0.1804, *Sirdavidia* = 0.1804, *Brieya* = 0.1804) (Figure 3). The CoMET analysis converged (ESS > 500) but we did not detect any significant mass extinction events (Figure 4E,F).

Table 1. Results from the RPANDA analyses. Abbreviations: speciation (λ); extinction (μ); corrected Akaike Information Criterion (AICc); the difference in AICc between the model with the lowest AICc and the others (ΔAICc); estimated speciation at present ($\lambda0$); rate of change in speciation rate (α, from the present to the past, negative rate meaning speciation has increased through time); estimated extinction at present ($\mu0$); rate of change in speciation rate (β, from the present to the past).

Models	Log Likelihood	AICc	ΔAICc	$\lambda0$	α	$\mu0$	β
Exponential λ (no μ)	−64.9652	134.531	0	0.224	−0.074	-	-
Constant λ/constant μ	−66.0258	136.651	2.120	0.241	-	0.200	-
Constant λ/exponential μ	−64.9436	137.150	2.619	0.198	-	0.057	0.008
Exponential λ/constant μ	−64.9652	137.194	2.663	0.224	−0.074	1.98×10^{-7}	-
Constant λ (no μ)	−68.1224	138.435	3.904	0.127	-	-	-
Exponential λ/exponential μ	−64.9652	140.153	5.622	0.224	−0.074	2.20×10^{-6}	0.023

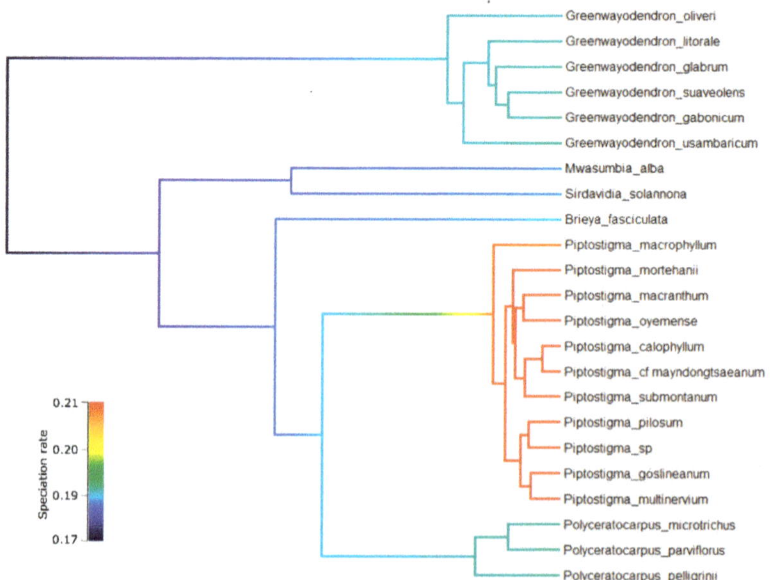

Figure 3. Results of the BAMM analysis. Phylogenetic tree of Piptostigmateae with inferred speciation rates on each branch (in events per lineage per million years).

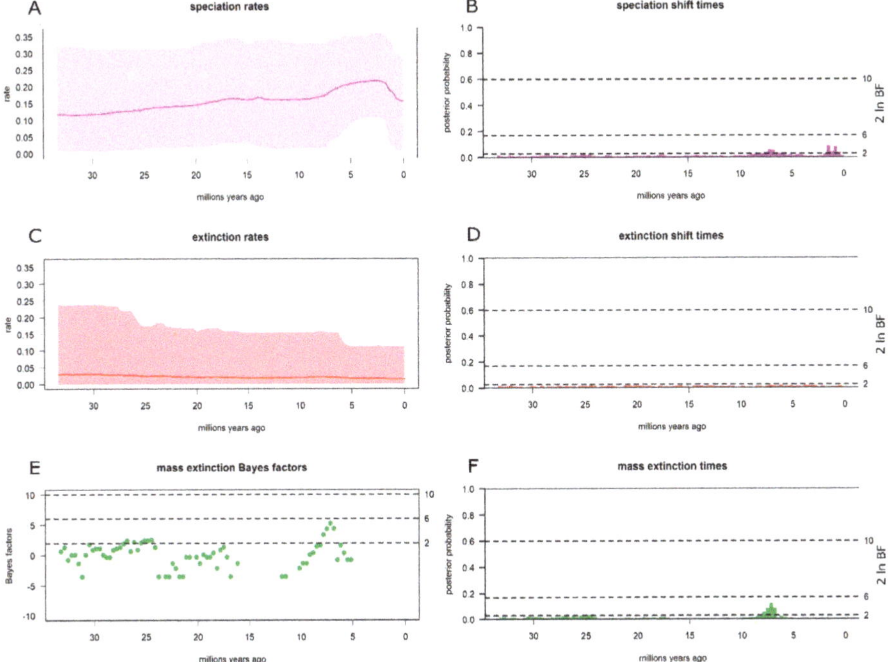

Figure 4. Results of the CoMET analyses. Inferred speciation (**A**) and extinction (**C**) rates through time (posterior mean and 95% credible interval), and possible mass extinction occurrence (**E**). Graph of the Bayes Factors (BF) testing models of speciation shift (**B**), extinction (**D**) and mass extinction (**F**) against null model. In total, 100 BF calculations were done from 33.35 Myr (Piptostigmateae crown node estimation) to present (≈1 BF estimation/0.33 Myr, represented by bars of posterior probability). A significant rate shift was indicated if 2 ln BF > 6.

4. Discussion

4.1. Temporal Estimates of African Rainforest Vicariance

In this study, we inferred a dated molecular phylogenetic tree of the Piptostigmateae and Annickieae tribes under a Bayesian inference method (BEAST, Figure 2). We used the same markers of Couvreur et al. (2019) [43], but we undertook a Bayesian phylogenetic analysis, which was not carried out in that paper (maximum likelihood only; a gene tree approach (ASTRAL) and concatenated approach using RAxML). Our Bayesian analysis, albeit based on 32 selected sequences, corroborated the results obtained using the gene trees species approach of Couvreur et al. (2019) [43] implemented in ASTRAL. Couvreur et al. (2019) [43] erected the Annickieae tribe based on the ASTRAL results, which is supported by our Bayesian analysis. Indeed, *Annickia* is inferred as a sister to the rest of the Malmeoideae with maximum support (Figure 2) (PP = 1).

Our dating analysis is the first time numerous nuclear markers were used to date clades within Annonaceae. Past studies have mainly relied on a few plastid markers [72,73] or full plastomes [74,75]. Interestingly, the 95% HPD interval of our molecular dating estimates of all the major clades of Annonaceae are consistent with previous global scale studies using BEAST but based on few plastid markers [41,76]. Nevertheless, the retrieved dates were overall more recent within the genus *Greenwayodendron* (Piptostigmateae) when compared to of the ages inferred by Migliore et al. [74] based on full plastome data and detailed intra-specific sampling. However, our age estimates do overlap based on the 95% HPD interval. We also note that, overall, our 95% HPD interval are smaller than the abovementioned studies. This can be explained by the use of more sequence data (32 exons/intron versus 10 plastid markers), which might also be more informative. Our larger dataset allowed us to select the markers most appropriate for divergence time estimation, making our estimates the most accurate and reliable for the tribe to date.

Our study provides two different time estimations of speciation events between species of ARFs endemic to the GC and EA regions, all occurring during the Neogene (Miocene/Pliocene). The oldest inferred split is between the two genera Mwasumbia (endemic to EA) and Sirdavidia (endemic to GC) dated to the Late Early Miocene (mean: 16.13 Myr). This date falls into the Middle Miocene Climatic Optimum (MMCO) (17–14.7 Myr) [77] and coincides with the stabilization of the East Antarctic Ice Sheet [78]. More importantly, it is followed by the Middle Miocene Climatic Transition (MMCT) (approximately 14.2 to 13.8 Myr). The MMCT was characterized by a global cooling and an aridification of the Congo basin and East Africa caused by the uplift of the East African plateau as well as other factors such as the East Tethys sea final closure [79–86]. In addition to these intra African events, other factors such as the stabilization of East Antarctic Ice Sheet could also have led to the increased aridity in Africa [78]. The expansion of grassland ecosystems and rainforest contraction have been identified during this period [2,33,79,87] (reviewed by [88]). Moreover, multiple independent molecular dating studies have also suggested vicariance events between forest restricted clades during this period in plants [24–26,89] including other African Annonaceae tribes [9], and animals such as forest-restricted chameleons [35]. Therefore, the divergence between Mwasumbia and Sirdavidia seems likely to have been the consequence of a vicariance event due to the aridification of the African continent. Some studies have questioned this middle Miocene ARF reconnection [90,91]. Linder [30] argued that East African vegetation before the East African uplift was more likely to be woodland. This was further supported by the lack of fossil evidence of a pan-African rainforest during the early/mid-Miocene. Nevertheless, the fossil records for these regions remain poor [30]. However, the concordance in timing of our results with previous dated molecular studies (e.g., [9]) lends support to an Early Miocene reconnection of the African rainforest block followed by the mid-Miocene fragmentation.

The two other vicariance events are temporally congruent, dated to the end of Miocene/beginning of the Pliocene (mean ages: 5.36 and 5.79 Myr). This supports the idea that a single event affected the divergence of species in both Greenwayodendron and Annickia at this time. During the second half of the Miocene, drier climatic conditions led to the expansion of savannas and grasslands, with C4 photosynthetic plants becoming progressively dominant, notably in EA [2,30]. Studies also report a progressive diversification of animal clades living in arid and open ecosystems during the late Miocene in Africa [19,92,93]. Suitable conditions for ARF expansion arose again only during the Early Pliocene (5–3.5 Myr), where the palynological record points to a moist climate, rapid diversification of rainforest taxa and retraction of savannas [2,13,81,88,94]. Our molecular dating also revealed that the ages of crown nodes of each of the genera *Polyceratocarpus*, *Piptostigma*, *Annickia* and *Greenwayodendron* match this period, which corroborates the possible re-expansion of ARF at this time. There is little evidence suggesting that ARFs expanded continuously again from East to West during the second half of Miocene. Thus, these two speciation events resulting from the vicariance of a continuous ARF at 5 Myr appears unlikely. As we recover a very similar vicariance in the two distinct clades, we suggest that two hypotheses could explain this pattern. Firstly, these two speciation events are the result of a same vicariance event due to a pan-African forest fragmentation, but later in geological times. In fact, abrupt climate changes are documented during the Late Pliocene–Early Pleistocene, with drying and cooling phases once again allowing the spread of grasses on the continent and causing extinction of rainforest taxa [2,30,88]. As indicated before, climatic conditions of the Early Pliocene could have reconnected ARFs, so maybe the speciation events we infer here occurred during this mid-Pliocene transition. Second, these events, despite their similarities, might be due to independent dispersal events linked to bird or other animal dispersal events. Indeed, both *Annickia* and *Greenwayodendron* contain some species that are large trees reaching up to 45 m, while other Piptostigmateae genera generally do not exceed 20 m [44,45,47,49]. In addition, fruits (monocarps) in *Annickia* and *Greenwayodendron* are small, clearly stalked and brightly colored appearing particularly adapted for dispersal by canopy dwelling birds or monkeys [47,49]. Finally, palynological data around 7 Myr indicate a possible forest expansion in both EA and GC [2,90], which could have reduced distances between rainforest blocks at this time, facilitating dispersal.

4.2. Piptostigmateae Diversification

Piptostigmateae started to diversify during the Paleogene around 36 Myr (Figure 2), just before the global and abrupt cooling event termed the Eocene–Oligocene Transition (EOT, 33 Myr). This event is suggested to have decreased tropical rainforest biodiversity worldwide [2,95] and is probably the cause of extinction in numerous plant groups, like African palms [96,97]. Palynological records show that the EOT led to an important turnover of biodiversity with the appearance of many current ARF clades during early Oligocene, such as the genus *Annona* [2]. However, the majority of Piptostigmateae species seem to have mainly diversified more recently, during the Pliocene, linked to improved climatic conditions for tropical rainforest taxa (see above) (Figure 2). Nonetheless, using BAMM, we did not detect significant speciation rate shifts at this time, which could be related to low statistical power linked to the low number of species known from the tribe and sampled here. Nevertheless, these results are in line with recent works on the Annonaceae family that found no evidence for diversification rate shifts for Piptostigmateae [98,99]. The BAMM analysis however also inferred a higher (but not significant) speciation rate for the genus *Piptostigma* (Figure 3). It remains unclear what might have led to this increase for this specific genus. *Piptostigma* is characterized within the tribe by the inner whorl of petals being much longer than the outer ones [46], a character only shared with the species poor genus *Brieya* (two species). All other genera in Piptostigmateae have longer outer petals than inner or are equal length [45]. Another distinctive character for *Piptostigma* is the presence of numerous secondary veins [46] with tight parallel tertiary venation (see Figure 1G) of the leaf blade (generally more than 20 pairs and up to 65 for some species, whereas other genera generally have less than 20 veins). High leaf vein density is suggested to increase photosynthetic and transpirational capacities [100,101] which could have conveyed a competitive advantage of *Piptostigma* over other Piptostigmateae genera, although this requires further testing.

Finally, because speciation within *Piptostigma* mainly took place during the Pleistocene (Figure 2), it cannot be excluded that diversification within this genus was linked to the lowland refuge hypothesis. This pattern, however, has rarely been reported in ARF restricted tree genera to date (e.g., *Carapa* [102]; *Coffea* [103]). In the latter genera, most species originated in the last 2 Myr. Most Pleistocene speciation events have been documented in herbaceous genera, such as *Begonias* or *Impatiens* (e.g., [7,104]), whereas tree genera have generally been inferred to originate before the Pleistocene [9,26,41,74,97,105,106].

The RPANDA analysis indicated that the speciation rate within Piptostigmateae increased exponentially through time, together with negligible extinction rates (Table 1). However, because we did not test any intermediate models between exponential and constant, the true speciation rate probably lines somewhere in between these two extremes. The negligible effect of extinction inferred on Piptostigmateae diversity is concordant with our BAMM analysis, which also found an increase of the speciation rate through time with no significant diversification rate shift (Figure 3). In addition, we did not detect any significant extinction shift, nor mass extinction events when using CoMET (Figure 4). Overall, our analyses do not detect a major role of extinction, either punctual or gradual, during the evolutionary history of Piptostigmateae. If interpreted as such, the low extinction rates inferred here are in line with a few other studies in the region. For example, in the palm subtribe Ancistrophyllinae, a clade of 22 climbing palm species (rattans) occurring across ARF, a near complete dated molecular phylogeny estimated low extinction rates [97]. They did, however, detect a signal of an ancient mass extinction event at the Oligocene–Eocene boundary, which was not detected here. In addition, a global study of palm diversification inferred overall low extinction rates for palms as a whole and did not detect a significant decrease in diversification rates for African genera [107]. These authors suggested that the "odd man out" pattern might not be the result of high extinction rates, but lower speciation rates in Africa. Interestingly, this hypothesis has received additional support from other pantropical plant families such as Sapotaceae [108] or Chrysobalanaceae [109] in which higher speciation rates and not extinction rates have been detected outside of Africa. Finally, for plants as a whole, a recent study found both higher speciation and extinction rates for the Neotropics, suggesting that higher diversity of the Neotropics is linked to higher turnover of taxa through time [110], but see [111]. Thus, these studies do not support the hypothesis that lower African rainforest diversity is linked to higher extinction rates [2,17,18]. However, we are aware of the difficulties of inferring extinction from molecular phylogenies without fossil data ([112], but also see [113]) and thus our results on extinction should be considered with caution.

5. Conclusions

Our dated molecular tree of both Piptostigmateae and Annickieae tribes detected two major vicariance events between central African and East African rainforest lineages: one during the Mid-Miocene, possibly linked to climatic and/or geological changes fragmenting the pre-existing pan-African rainforest at this time; the second one is estimated to have occurred at the beginning of the Pliocene, and is harder to link to a rainforest vicariance event. Our results show that Piptostigmateae species mainly diversified during the Pliocene and Pleistocene. The genus *Piptostigma*, with 13 species, is one of the rare examples of tree clades diversifying mostly during the Pleistocene. We did not detect high levels of overall extinction and/or mass extinction events to explain present-day diversity. This suggests a possible low impact of extinction of the evolutionary history of this clade. Therefore, our results are in line with other studies detecting low extinction rates within some African clades. In addition, Piptostigmateae was not associated with significant shifts in diversification rates. Indeed, the best fitting model suggested that the speciation rate of Piptostigmateae increased exponentially over time. These results imply that the global diversity of this tribe was not impacted by major abiotic changes during the Neogene. Overall, our study adds to the knowledge of ARF evolution and the patterns of diversification that have occurred in this species-rich ecosystem.

Supplementary Materials: The following are available online at http://www.mdpi.com/1424-2818/12/6/227/s1, Table S1: List and distribution of the 30 species used in this study (23 from the Piptostigmateae and 7 from the Annickieae) used for the phylogenetic reconstruction, with their main floristic bioregions as identified by Droissart et al. (2018) [114]. Grey highlights strict East African endemic species. All the others are endemics of the Guineo–Congolian region. Table S2: Specimen information used for this study with associated sequencing statistics. Total reads: Total number of reads recovered for each barcoded tag; mapped reads: number of reads mapped to reference exonic sequences; % enrichment: percentage of reads correctly mapped to reference exonic sequences. 10x coverage: proportion of the reference sequences covered with at least 10 base pairs; mean depth: average number of reads covering reference sequences. Table S3: Table listing the 32 selected sequences chosen for the BEAST analysis, with their best nucleotide substitution models selected with the BIC criterion and their tree length with root-to-tip variance. Table S4: Age estimate of the extant taxa for the six Piptostigmateae genera and Annickia using BEAST. The crown node estimation of Brieya was not possible because we did not include the second species of this genus. Table S5: Results of the TESS analysis. ESS for all MCMC where above 300 except for the Exponential λ/exponential μ model (less than 100). Abbreviations: λ: speciation; μ: extinction; $\lambda0$: estimated speciation at present; α: rate of change in speciation; $\mu0$: estimated extinction at present; β: rate of change in speciation rate (from the present to the past). Table S6: Output of the computeBayesFactors command from BAMMtools. Bayes Factors values for the model with k shifts relative to the model with 0 rate shifts. Table S7: Results of the Nested Sampling model selection.

Author Contributions: Conceptualization, T.L.P.C. and A.J.H.; methodology, A.J.H., K.B.; validation, K.B., A.J.H. and T.L.P.C.; formal analysis, B.B. and A.J.H.; resources, T.L.P.C. and J.-P.G.; data curation, J.-P.G. and K.B.; writing—original draft preparation, B.B. and T.L.P.C.; writing—review and editing, A.J.H., B.S. and J.-P.G.; visualization, B.B. and T.L.P.C.; supervision, B.S. and T.L.P.C.; project administration, T.L.P.C.; funding acquisition, T.L.P.C. All authors have read and agreed to the published version of the manuscript.

Funding: This research received no external funding.

Acknowledgments: We thank Raoul Niangadouma and Narcisse Kamdem for help in the field. The authors acknowledge the IRD itrop HPC (South Green Platform) at IRD Montpellier for providing HPC resources. This study was supported by the Agence Nationale de la Recherche (grant number ANR-15-CE02-0002-01 to TLPC). We are grateful to the Centre National de la Recherche Scientifique et Technique (CENAREST), the Agence National des Parqués Nationaux (ANPN) and Prof. Bourobou Bourobou for research permits (AR0020/16; AR0036/15

(CENAREST) and AE16014 (ANPN)). Fieldwork in Cameroon was undertaken under the "accord cadre de cooperation" between the IRD and Ministère de la Recherche Scientifique et Technique (MINRESI). Finally, we thank three anonymous reviewers for providing detailed comments on a previous version of this article.

Conflicts of Interest: The authors declare no conflict of interest.

References

1. Whitmore, T.C. *An Introduction to Tropical Rain Forests*; Clarendon Press: Oxford, UK, 1998; ISBN 978-0-19-850147-3.
2. Morley, R.J. *Origin and Evolution of Tropical Rain Forests*; John Wiley & Sons: New York, NY, USA, 2000; ISBN 978-0-471-98326-2.
3. Linder, H.P. Plant diversity and endemism in sub-Saharan tropical Africa. *J. Biogeogr.* **2001**, *28*, 169–182. [CrossRef]
4. Klopper, R.R.; Gautier, L.; Chatelain, C.; Smith, G.F.; Spichiger, R. Floristics of the Angiosperm flora of Sub-Saharan Africa: An analysis of the African plant checklist and database. *Taxon* **2007**, *56*, 201–208. [CrossRef]
5. Sosef, M.S.M.; Dauby, G.; Blach-Overgaard, A.; van der Burgt, X.; Catarino, L.; Damen, T.; Deblauwe, V.; Dessein, S.; Dransfield, J.; Droissart, V.; et al. Exploring the floristic diversity of tropical Africa. *BMC Biol.* **2017**, *15*, 15. [CrossRef] [PubMed]
6. Linder, H.P. The evolution of African plant diversity. *Front. Ecol. Evol.* **2014**, *2*, 38. [CrossRef]
7. Plana, V.; Gascoigne, A.; Forrest, L.L.; Harris, D.; Pennington, R.T. Pleistocene and pre-Pleistocene Begonia speciation in Africa. *Mol. Phylogenet. Evol.* **2004**, *31*, 449–461. [CrossRef] [PubMed]
8. Auvrey, G.; Harris, D.J.; Richardson, J.E.; Newman, M.F.; Särkinen, T.E. Phylogeny and dating of Aframomum (Zingiberaceae). In *Diversity, Phylogeny, and Evolution in the Monocotyledons*; Seberg, O., Peterson, G., Barfod, A., Davis, J.I., Eds.; Aarhus University Press: Aarhus, Denmark, 2010; pp. 287–305. ISBN 978-87-7934-398-6.
9. Couvreur, T.L.; Chatrou, L.W.; Sosef, M.S.; Richardson, J.E. Molecular phylogenetics reveal multiple tertiary vicariance origins of the African rain forest trees. *BMC Biol.* **2008**, *6*, 54. [CrossRef] [PubMed]
10. Couvreur, T.L.P.; Porter-Morgan, H.; Wieringa, J.J.; Chatrou, L.W. Little ecological divergence associated with speciation in two African rain forest tree genera. *BMC Evol. Biol.* **2011**, *11*, 296. [CrossRef] [PubMed]
11. Sosef, M.S.M. Refuge Begonias: Taxonomy, phylogeny and historical biogeography of Begonia sect. Loasibegonia and sect. Scutobegonia in relation to glacial rain forest refuges in Africa. Studies in Begonia V. *Wagening. Agric. Univ. Pap.* **1994**, *94*, 1–306.
12. Maley, J.; Brenac, P. Vegetation dynamics, palaeoenvironments and climatic changes in the forests of western Cameroon during the last 28,000 years B.P. *Rev. Palaeobot. Palynol.* **1998**, *99*, 157–187. [CrossRef]
13. Plana, V. Mechanisms and tempo of evolution in the African Guineo–Congolian rainforest. *Philos. Trans. R. Soc. Lond. B Biol. Sci.* **2004**, *359*, 1585–1594. [CrossRef]
14. Hardy, O.J.; Born, C.; Budde, K.; Daïnou, K.; Dauby, G.; Duminil, J.; Ewédjé, E.-E.B.; Gomez, C.; Heuertz, M.; Koffi, G.K.; et al. Comparative phylogeography of African rain forest trees: A review of genetic signatures of vegetation history in the Guineo-Congolian region. *Comptes Rendus Geosci.* **2013**, *345*, 284–296. [CrossRef]
15. Richards, P.W. Africa, the "Odd man out". In *Tropical Forest Ecosystems of Africa and South America: A Comparative Review*; Meggers, B.J., Ayensu, E.S., Duckworth, W.D., Eds.; Smithsonian Institution Press: Washington, DC, USA, 1973; ISBN 978-0-87474-125-4.
16. Couvreur, T.L.P. Odd man out: Why are there fewer plant species in African rain forests? *Plant Syst. Evol.* **2015**, *301*, 1299–1313. [CrossRef]
17. Axelrod, D.I.; Raven, P.H. Late Cretaceous and Tertiary vegetation history of Africa. In *Biogeography and Ecology of Southern Africa*; Werger, M.J.A., Ed.; W. Junk bv Publishers: The Hague, The Netherlands, 1978; pp. 77–130. ISBN 978-94-009-9951-0.
18. Richards, P.W. *The Tropical Rain Forest: An Ecological Study*, 2nd ed.; Cambridge University Press: Cambridge, UK, 1996; ISBN 978-0-521-42194-2.
19. Bobe, R. The evolution of arid ecosystems in eastern Africa. *J. Arid Environ.* **2006**, *66*, 564–584. [CrossRef]
20. White, F. The Guineo-Congolian region and its relationships to other phytochoria. *Bull. Jard. Bot. Natl. Belg.* **1979**, *49*, 11–55. [CrossRef]

21. Wasser, S.K.; Lovett, J.C. *Biogeography and Ecology of the Rainforests of Eastern Africa*; Lovett, J.C., Wasser, S.K., Eds.; Cambridge University Press: Cambridge, UK, 1993; ISBN 978-0-521-06898-7.

22. Burgess, N.D.; Clarke, G.P.; Rodgers, W.A. Coastal forests of eastern Africa: Status, endemism patterns and their potential causes. *Biol. J. Linn. Soc.* **1998**, *64*, 337–367. [CrossRef]

23. Burgess, N.D.; Clarke, G.P. *Coastal Forests of Eastern Africa*; IUCN - The World Conservation Union, Publications Services Unit: Cambridge, UK, 2000; ISBN 978-2-8317-0436-4.

24. Dimitrov, D.; Nogués-Bravo, D.; Scharff, N. Why Do Tropical Mountains Support Exceptionally High Biodiversity? The Eastern Arc Mountains and the Drivers of Saintpaulia Diversity. *PLoS ONE* **2012**, *7*, e48908. [CrossRef] [PubMed]

25. Pokorny, L.; Riina, R.; Mairal, M.; Meseguer, A.S.; Culshaw, V.; Cendoya, J.; Serrano, M.; Carbajal, R.; Ortiz, S.; Heuertz, M.; et al. Living on the edge: Timing of Rand Flora disjunctions congruent with ongoing aridification in Africa. *Front. Genet.* **2015**, *6*, 154. [CrossRef]

26. Tosso, F.; Hardy, O.J.; Doucet, J.-L.; Daïnou, K.; Kaymak, E.; Migliore, J. Evolution in the Amphi-Atlantic tropical genus Guibourtia (Fabaceae, Detarioideae), combining NGS phylogeny and morphology. *Mol. Phylogenet. Evol.* **2018**, *120*, 83–93. [CrossRef]

27. Lovett, J.C. Climatic history and forest distribution in eastern Africa. In *Biogeography and Ecology of the Rainforests of Eastern Africa*; Lovett, J.C., Wasser, S.K., Eds.; Cambridge University Press: Cambridge, UK, 1993; pp. 23–29. ISBN 0-521-43083-6.

28. Morley, R.J. Cretaceous and Tertiary climate change and the past distribution of megathermal rainforests. In *Tropical Rainforest Responses to Climatic Change*; Bush, M.B., Flenley, J., Eds.; Praxis Publishing: Chichester, UK, 2007; pp. 1–31. ISBN 978-3-642-05382-5.

29. Willis, K.; McElwain, J. *The Evolution of Plants*, 2nd ed.; Oxford University Press: Oxford, UK, 2014; ISBN 978-0-19-929223-3.

30. Linder, H.P. East African Cenozoic vegetation history. *Evol. Anthropol. Issues News Rev.* **2017**, *26*, 300–312. [CrossRef]

31. Davis, C.C.; Bell, C.D.; Fritsch, P.W.; Mathews, S. Phylogeny of Acridocarpus-Brachylophon (Malpighiaceae): Implications for Tertiary tropical floras and Afroasian biogeography. *Evolution* **2002**, *56*, 2395–2405. [CrossRef]

32. Coetzee, J.A. African flora since the terminal Jurassic. In *Biological Relationships between Africa and South America*; Goldblatt, P., Ed.; Yale University Press: New Haven, CT, USA, 1993; pp. 37–61. ISBN 978-0-300-05375-3.

33. Jacobs, B.F.; Kingston, J.D.; Jacobs, L.L. The Origin of Grass-Dominated Ecosystems. *Ann. Mo. Bot. Gard.* **1999**, *86*, 590–643. [CrossRef]

34. Voelker, G.; Outlaw, R.K.; Bowie, R.C.K. Pliocene forest dynamics as a primary driver of African bird speciation: African forest refugia. *Glob. Ecol. Biogeogr.* **2010**, *19*, 111–121. [CrossRef]

35. Tolley, K.A.; Townsend, T.M.; Vences, M. Large-scale phylogeny of chameleons suggests African origins and Eocene diversification. *Proc. R. Soc. B Biol. Sci.* **2013**, *280*, 20130184. [CrossRef]

36. Loader, S.P.; Pisani, D.; Cotton, J.A.; Gower, D.J.; Day, J.J.; Wilkinson, M. Relative time scales reveal multiple origins of parallel disjunct distributions of African caecilian amphibians. *Biol. Lett.* **2007**, *3*, 505–508. [CrossRef]

37. Chatrou, L.W.; Pirie, M.D.; Erkens, R.H.J.; Couvreur, T.L.P.; Neubig, K.M.; Abbott, J.R.; Mols, J.B.; Maas, J.W.; Saunders, R.M.K.; Chase, M.W. A new subfamilial and tribal classification of the pantropical flowering plant family Annonaceae informed by molecular phylogenetics. *Bot. J. Linn. Soc.* **2012**, *169*, 5–40. [CrossRef]

38. Guo, X.; Tang, C.C.; Thomas, D.C.; Couvreur, T.L.P.; Saunders, R.M.K. A mega-phylogeny of the Annonaceae: Taxonomic placement of five enigmatic genera and support for a new tribe, Phoenicantheae. *Sci. Rep.* **2017**, *7*, 7323. [CrossRef] [PubMed]

39. Sonké, B.; Couvreur, T.L. Tree diversity of the Dja Faunal Reserve, southeastern Cameroon. *Biodivers. Data J.* **2014**, e1049. [CrossRef]

40. Doyle, J.A.; Le Thomas, A. Cladistic analysis and pollen evolution in Annonaceae. *Acta Bot. Gallica* **1994**, *141*, 149–170. [CrossRef]

41. Couvreur, T.L.P.; Forest, F.; Baker, W.J. Origin and global diversification patterns of tropical rain forests: Inferences from a complete genus-level phylogeny of palms. *BMC Biol.* **2011**, *9*, 44. [CrossRef]

42. Thomas, D.C.; Tang, C.C.; Saunders, R.M.K. Historical biogeography of *Goniothalamus* and Annonaceae tribe Annoneae: Dispersal-vicariance patterns in tropical Asia and intercontinental tropical disjunctions revisited. *J. Biogeogr.* **2017**, *44*, 2862–2876. [CrossRef]

43. Couvreur, T.L.P.; Helmstetter, A.J.; Koenen, E.J.M.; Brandão, R.D.; Little, S.; Sauquet, H.; Erkens, R.H.J. Phylogenomics of the major tropical plant family Annonaceae using targeted enrichment of nuclear genes. *Front. Plant Sci.* **2019**, 1941. [CrossRef] [PubMed]

44. Couvreur, T.L.P.; van der Ham, R.W.J.M.; Mbele, Y.M.; Mbago, F.M.; Johnson, D.M. Molecular and Morphological Characterization of a New Monotypic Genus of Annonaceae, Mwasumbia from Tanzania. *Syst. Bot.* **2009**, *34*, 266–276. [CrossRef]

45. Couvreur, T.L.P.; Niangadouma, R.; Sonké, B.; Sauquet, H. Sirdavidia, an extraordinary new genus of Annonaceae from Gabon. *PhytoKeys* **2015**, 1–19. [CrossRef] [PubMed]

46. Ghogue, J.-P.; Sonké, B.; Couvreur, T.L.P. Taxonomic revision of the African genera Brieya and Piptostigma (Annonaceae). *Plant Ecol. Evol.* **2017**, *150*, 173–216. [CrossRef]

47. Lissambou, B.-J.; Hardy, O.J.; Atteke, C.; Stevart, T.; Dauby, G.; Mbatchi, B.; Sonke, B.; Couvreur, T.L.P. Taxonomic revision of the African genus Greenwayodendron (Annonaceae). *PhytoKeys* **2018**, 55–93. [CrossRef]

48. Marshall, A.R.; Couvreur, T.L.P.; Summers, A.L.; Deere, N.J.; Luke, W.R.Q.; Ndangalasi, H.J.; Sparrow, S.; Johnson, D.M. A new species in the tree genus Polyceratocarpus (Annonaceae) from the Udzungwa Mountains of Tanzania. *PhytoKeys* **2016**, 63–76. [CrossRef]

49. Versteegh, C.P.C.; Sosef, M.S.M. Revision of the African genus Annickia (Annonaceae). *Syst. Geograph. Plants* **2007**, 91–118. [CrossRef]

50. Johnson, M.G.; Gardner, E.M.; Liu, Y.; Medina, R.; Goffinet, B.; Shaw, A.J.; Zerega, N.J.C.; Wickett, N.J. HybPiper: Extracting Coding Sequence and Introns for Phylogenetics from High-Throughput Sequencing Reads Using Target Enrichment. *Appl. Plant Sci.* **2016**, *4*, 1600016. [CrossRef]

51. Katoh, K.; Standley, D.M. MAFFT Multiple Sequence Alignment Software Version 7: Improvements in Performance and Usability. *Mol. Biol. Evol.* **2013**, *30*, 772–780. [CrossRef]

52. Castresana, J. Selection of Conserved Blocks from Multiple Alignments for Their Use in Phylogenetic Analysis. *Mol. Biol. Evol.* **2000**, *17*, 540–552. [CrossRef]

53. Stamatakis, A. RAxML version 8: A tool for phylogenetic analysis and post-analysis of large phylogenies. *Bioinformatics* **2014**, *30*, 1312–1313. [CrossRef] [PubMed]

54. Mai, U., Mirarab, S. TreeShrink: Fast and accurate detection of outlier long branches in collections of phylogenetic trees. *BMC Genom.* **2018**, *19*, 272. [CrossRef] [PubMed]

55. Smith, S.A.; Brown, J.W.; Walker, J.F. So many genes, so little time: A practical approach to divergence-time estimation in the genomic era. *PLoS ONE* **2018**, *13*, e0197433. [CrossRef] [PubMed]

56. Darriba, D.; Posada, D.; Kozlov, A.M.; Stamatakis, A.; Morel, B.; Flouri, T. ModelTest-NG: A new and scalable tool for the selection of DNA and protein evolutionary models. *bioRxiv* **2009**, 612903. [CrossRef]

57. Drummond, A.J.; Suchard, M.A.; Xie, D.; Rambaut, A. Bayesian Phylogenetics with BEAUti and the BEAST 1.7. *Mol. Biol. Evol.* **2012**, *29*, 1969–1973. [CrossRef]

58. Pirie, M.D.; Doyle, J.A. Dating clades with fossils and molecules: The case of Annonaceae: Dating clades in Annonaceae. *Bot. J. Linn. Soc.* **2012**, *169*, 84–116. [CrossRef]

59. Russel, P.M.; Brewer, B.J.; Klaere, S.; Bouckaert, R.R. Model Selection and Parameter Inference in Phylogenetics Using Nested Sampling. *Syst. Biol.* **2018**, *68*, 219–233. [CrossRef]

60. Rambaut, A.; Drummond, A.J.; Xie, D.; Baele, G.; Suchard, M.A. Posterior Summarization in Bayesian Phylogenetics Using Tracer 1.7. *Syst. Biol.* **2018**, *67*, 901–904. [CrossRef]

61. Bouckaert, R.; Heled, J.; Kühnert, D.; Vaughan, T.; Wu, C.-H.; Xie, D.; Suchard, M.A.; Rambaut, A.; Drummond, A.J. BEAST 2: A Software Platform for Bayesian Evolutionary Analysis. *PLoS Comput. Biol.* **2014**, *10*, e1003537. [CrossRef]

62. Paradis, E.; Claude, J.; Strimmer, K. APE: Analyses of Phylogenetics and Evolution in R language. *Bioinformatics* **2004**, *20*, 289–290. [CrossRef]

63. R Core Team. *R: A Language and Environment for Statistical Computing*; R Foundation for Statistical Computing: Vienna, Austria, 2019. Available online: http://www.R-project.org (accessed on 15 May 2019).

64. Morlon, H.; Lewitus, E.; Condamine, F.L.; Manceau, M.; Clavel, J.; Drury, J. RPANDA: An R package for macroevolutionary analyses on phylogenetic trees. *Methods Ecol. Evol.* **2016**, *7*, 589–597. [CrossRef]

65. Burnham, K.P.; Anderson, D.R. *Model Selection and Multimodel Inference: A Practical Information-Theoretic Approach*, 2nd ed.; Springer: New York, NY, USA, 2002; ISBN 978-0-387-22456-5.

66. Höhna, S.; May, M.R.; Moore, B.R. TESS: An R package for efficiently simulating phylogenetic trees and performing Bayesian inference of lineage diversification rates. *Bioinformatics* **2016**, *32*, 789–791. [CrossRef] [PubMed]

67. Rabosky, D.L. Automatic Detection of Key Innovations, Rate Shifts, and Diversity-Dependence on Phylogenetic Trees. *PLoS ONE* **2014**, *9*, e89543. [CrossRef] [PubMed]

68. Rabosky, D.L.; Mitchell, J.S.; Chang, J. Is BAMM Flawed? Theoretical and Practical Concerns in the Analysis of Multi-Rate Diversification Models. *Syst. Biol.* **2017**, *66*, 477–498. [CrossRef]

69. Plummer, M.; Best, N.; Cowles, K.; Vines, K. CODA: Convergence diagnosis and output analysis for MCMC. *R News* **2006**, *6*, 7–11.

70. Culshaw, V.; Stadler, T.; Sanmartín, I. Exploring the power of bayesian birth-death skyline models to detect mass extinction events from phylogenies with only extant taxa. *Evolution* **2019**, *73*, 1133–1150. [CrossRef]

71. Kass, R.E.; Raftery, A.E. Bayes Factors. *J. Am. Stat. Assoc.* **1995**, *90*, 773–795. [CrossRef]

72. Richardson, J.E.; Chatrou, L.W.; Mols, J.B.; Erkens, R.H.J.; Pirie, M.D. Historical biogeography of two cosmopolitan families of flowering plants: Annonaceae and Rhamnaceae. *Philos. Trans. R. Soc. Lond. B Ser.* **2004**, *359*, 1495–1508. [CrossRef]

73. Xue, B.; Guo, X.; Landis, J.B.; Sun, M.; Tang, C.C.; Soltis, P.S.; Soltis, D.E.; Saunders, R.M.K. Accelerated diversification correlated with functional traits shapes extant diversity of the early divergent angiosperm family Annonaceae. *Mol. Phylogenet. Evol.* **2020**, *142*, 106659. [CrossRef]

74. Migliore, J.; Kaymak, E.; Mariac, C.; Couvreur, T.L.P.; Lissambou, B.; Piñeiro, R.; Hardy, O.J. Pre-Pleistocene origin of phylogeographical breaks in African rain forest trees: New insights from *Greenwayodendron* (Annonaceae) phylogenomics. *J. Biogeogr.* **2018**, *46*, 212–223. [CrossRef]

75. Lopes, J.C.; Chatrou, L.W.; Mello-Silva, R.; Rudall, P.J.; Sajo, M.G. Phylogenomics and evolution of floral traits in the Neotropical tribe Malmeeae (Annonaceae). *Mol. Phylogenet. Evol.* **2018**, *118*, 379–391. [CrossRef] [PubMed]

76. Thomas, D.C.; Chatrou, L.W.; Stull, G.W.; Johnson, D.M.; Harris, D.J.; Thongpairoj, U.; Saunders, R.M.K. The historical origins of palaeotropical intercontinental disjunctions in the pantropical flowering plant family Annonaceae. *Perspect. Plant Ecol. Evol. Syst.* **2015**, *17*, 1–16. [CrossRef]

77. Shevenell, A.E.; Kennett, J.P.; Lea, D.W. Middle Miocene ice sheet dynamics, deep-sea temperatures, and carbon cycling: A Southern Ocean perspective. *Geochem. Geophys. Geosyst.* **2008**, *9*. [CrossRef]

78. Flower, B.P.; Kennett, J.P. The middle Miocene climatic transition: East Antarctic ice sheet development, deep ocean circulation and global carbon cycling. *Palaeogeogr. Palaeoclimatol. Palaeoecol.* **1994**, *108*, 537–555. [CrossRef]

79. Jacobs, B.F. Palaeobotanical studies from tropical Africa: Relevance to the evolution of forest, woodland and savannah biomes. *Philos. Trans. R. Soc. Lond. B. Biol. Sci.* **2004**, *359*, 1573–1583. [CrossRef]

80. Sepulchre, P.; Ramstein, G.; Fluteau, F.; Schuster, M.; Tiercelin, J.-J.; Brunet, M. Tectonic Uplift and Eastern Africa Aridification. *Science* **2006**, *313*, 1419–1423. [CrossRef]

81. Senut, B.; Pickford, M.; Ségalen, L. Neogene desertification of Africa. *Comptes Rendus Geosci.* **2009**, *341*, 591–602. [CrossRef]

82. Sepulchre, P.; Ramstein, G.; Schuster, M. Modelling the impact of tectonics, surface conditions and sea surface temperatures on Saharan and sub-Saharan climate evolution. *Comptes Rendus Geosci.* **2009**, *341*, 612–620. [CrossRef]

83. Prömmel, K.; Cubasch, U.; Kaspar, F. A regional climate model study of the impact of tectonic and orbital forcing on African precipitation and vegetation. *Palaeogeogr. Palaeoclimatol. Palaeoecol.* **2013**, *369*, 154–162. [CrossRef]

84. Sommerfeld, A.; Prömmel, K.; Cubasch, U. The East African Rift System and the impact of orographic changes on regional climate and the resulting aridification. *Int. J. Earth Sci.* **2016**, *105*, 1779–1794. [CrossRef]

85. Moucha, R.; Forte, A.M. Changes in African topography driven by mantle convection. *Nat. Geosci.* **2011**, *4*, 707–712. [CrossRef]

86. Hamon, N.; Sepulchre, P.; Lefebvre, V.; Ramstein, G. The role of eastern Tethys seaway closure in the Middle Miocene Climatic Transition (ca. 14 Ma). *Clim. Past* **2013**, *9*, 2687–2702. [CrossRef]

87. Retallack, G.J.; Dugas, D.P.; Bestland, E.A. Fossil Soils and Grasses of a Middle Miocene East-African Grassland. *Science* **1990**, *247*, 1325–1328. [CrossRef]

88. Jacobs, B.F.; Pan, A.D.; Scotese, C.R. A review of the Cenozoic vegetation history of Africa. In *Cenozoic Mammals of Africa*; Werdelin, L., Sanders, J., Eds.; University of California Press: Berkeley, CA, USA, 2010; pp. 57–72. ISBN 978-0-520-25721-4.

89. Mankga, L.T.; Yessoufou, K.; Chitakira, M. On the origin and diversification history of the African genus Encephalartos. *S. Afr. J. Bot.* **2020**, *130*, 231–239. [CrossRef]

90. Bonnefille, R. Cenozoic vegetation, climate changes and hominid evolution in tropical Africa. *Glob. Planet. Chang.* **2010**, *72*, 390–411. [CrossRef]

91. Fer, I.; Tietjen, B.; Jeltsch, F.; Trauth, M.H. Modelling vegetation change during Late Cenozoic uplift of the East African plateaus. *Palaeogeogr. Palaeoclimatol. Palaeoecol.* **2017**, *467*, 120–130. [CrossRef]

92. Hassanin, A.; Delsuc, F.; Ropiquet, A.; Hammer, C.; Jansen van Vuuren, B.; Matthee, C.; Ruiz-Garcia, M.; Catzeflis, F.; Areskoug, V.; Nguyen, T.T.; et al. Pattern and timing of diversification of Cetartiodactyla (Mammalia, Laurasiatheria), as revealed by a comprehensive analysis of mitochondrial genomes. *C. R. Biol.* **2012**, *335*, 32–50. [CrossRef]

93. Portillo, F.; Branch, W.R.; Conradie, W.; Rödel, M.-O.; Penner, J.; Barej, M.F.; Kusamba, C.; Muninga, W.M.; Aristote, M.M.; Bauer, A.M.; et al. Phylogeny and biogeography of the African burrowing snake subfamily Aparallactinae (Squamata: Lamprophiidae). *Mol. Phylogenet. Evol.* **2018**, *127*, 288–303. [CrossRef]

94. Pickford, M.; Senut, B.; Mourer-Chauviré, C. Early Pliocene Tragulidae and peafowls in the Rift Valley, Kenya: Evidence for rainforest in East Africa. *Comptes Rendus Palevol* **2004**, *3*, 179–189. [CrossRef]

95. Jaramillo, C.; Rueda, M.J.; Mora, G. Cenozoic Plant Diversity in the Neotropics. *Science* **2006**, *311*, 1893–1896. [CrossRef]

96. Pan, A.D.; Jacobs, B.F.; Dransfield, J.; Baker, W.J. The fossil history of palms (Arecaceae) in Africa and new records from the Late Oligocene (28-27 Mya) of north-western Ethiopia. *Bot. J. Linn. Soc.* **2006**, *151*, 69–81. [CrossRef]

97. Faye, A.; Pintaud, J.-C.; Baker, W.J.; Vigouroux, Y.; Sonke, B.; Couvreur, T.L.P. Phylogenetics and diversification history of African rattans (Calamoideae, Ancistrophyllinae). *Bot. J. Linn. Soc.* **2016**, *182*, 256–271. [CrossRef]

98. Erkens, R.H.J.; Mennega, E.A.; Westra, L.Y.T. A concise bibliographic overview of Annonaceae. *Bot. J. Linn. Soc.* **2012**, *169*, 41–73. [CrossRef]

99. Massoni, J.; Couvreur, T.L.; Sauquet, H. Five major shifts of diversification through the long evolutionary history of Magnoliidae (angiosperms). *BMC Evol. Biol.* **2015**, *15*, 49. [CrossRef]

100. Boyce, C.K.; Brodribb, T.J.; Feild, T.S.; Zwieniecki, M.A. Angiosperm leaf vein evolution was physiologically and environmentally transformative. *Proc. Biol. Sci.* **2009**, *276*, 1771–1776. [CrossRef] [PubMed]

101. Brodribb, T.J.; Feild, T.S. Leaf hydraulic evolution led a surge in leaf photosynthetic capacity during early angiosperm diversification. *Ecol. Lett.* **2010**, *13*, 175–183. [CrossRef]

102. Koenen, E.J.M.; Clarkson, J.J.; Pennington, T.D.; Chatrou, L.W. Recently evolved diversity and convergent radiations of rainforest mahoganies (Meliaceae) shed new light on the origins of rainforest hyperdiversity. *New Phytol.* **2015**, *207*, 327–339. [CrossRef] [PubMed]

103. Kainulainen, K.; Razafimandimbison, S.G.; Wikström, N.; Bremer, B. Island hopping, long-distance dispersal and species radiation in the Western Indian Ocean: Historical biogeography of the Coffeeae alliance (Rubiaceae). *J. Biogeogr.* **2017**, *44*, 1966–1979. [CrossRef]

104. Janssens, S.B.; Knox, E.B.; Huysmans, S.; Smets, E.F.; Merckx, V.S.F.T. Rapid radiation of Impatiens (Balsaminaceae) during Pliocene and Pleistocene: Result of a global climate change. *Mol. Phylogenet. Evol.* **2009**, *52*, 806–824. [CrossRef]

105. Faye, A.; Deblauwe, V.; Mariac, C.; Richard, D.; Sonké, B.; Vigouroux, Y.; Couvreur, T.L.P. Phylogeography of the genus Podococcus (Palmae/Arecaceae) in Central African rain forests: Climate stability predicts unique genetic diversity. *Mol. Phylogenet. Evol.* **2016**, *105*, 126–138. [CrossRef]

106. Monthe, F.K.; Migliore, J.; Duminil, J.; Bouka, G.; Demenou, B.B.; Doumenge, C.; Blanc-Jolivet, C.; Ekué, M.R.M.; Hardy, O.J. Phylogenetic relationships in two African Cedreloideae tree genera (Meliaceae) reveal multiple rain/dry forest transitions. *Perspect. Plant Ecol. Evol. Syst.* **2019**, *37*, 1–10. [CrossRef]

107. Baker, W.J.; Couvreur, T.L.P. Global biogeography and diversification of palms sheds light on the evolution of tropical lineages. I. Historical biogeography. *J. Biogeogr.* **2013**, *40*, 274–285. [CrossRef]

108. Armstrong, K.E.; Stone, G.N.; Nicholls, J.A.; Valderrama, E.; Anderberg, A.A.; Smedmark, J.; Gautier, L.; Naciri, Y.; Milne, R.; Richardson, J.E. Patterns of diversification amongst tropical regions compared: A case study in Sapotaceae. *Front. Genet.* **2014**, *5*, 362. [CrossRef] [PubMed]

109. Bardon, L.; Chamagne, J.; Dexter, K.G.; Sothers, C.A.; Prance, G.T.; Chave, J. Origin and evolution of Chrysobalanaceae: Insights into the evolution of plants in the Neotropics. *Bot. J. Linn. Soc.* **2013**, *171*, 19–37. [CrossRef]

110. Antonelli, A.; Zizka, A.; Silvestro, D.; Scharn, R.; Cascales-Miñana, B.; Bacon, C.D. An engine for global plant diversity: Highest evolutionary turnover and emigration in the American tropics. *Front. Genet.* **2015**, *6*, 130. [CrossRef]

111. de la Estrella, M.; Forest, F.; Wieringa, J.J.; Fougère-Danezan, M.; Bruneau, A. Insights on the evolutionary origin of Detarioideae, a clade of ecologically dominant tropical African trees. *New Phytol.* **2017**, *214*, 1722–1735. [CrossRef] [PubMed]

112. Rabosky, D.L. Extinction Rates Should Not Be Estimated from Molecular Phylogenies. *Evolution* **2010**, *64*, 1816–1824. [CrossRef]

113. Beaulieu, J.M.; O'Meara, B.C. Extinction can be estimated from moderately sized molecular phylogenies. *Evolution* **2015**, *69*, 1036–1043. [CrossRef]

114. Droissart, V.; Dauby, G.; Hardy, O.J.; Deblauwe, V.; Harris, D.J.; Janssens, S.; Mackinder, B.A.; Blach-Overgaard, A.; Sonké, B.; Sosef, M.S.M.; et al. Beyond trees: Biogeographical regionalization of tropical Africa. *J. Biogeogr.* **2018**, *45*, 1153–1167. [CrossRef]

Article

Phylogenomic Study of *Monechma* Reveals Two Divergent Plant Lineages of Ecological Importance in the African Savanna and Succulent Biomes

Iain Darbyshire [1,*,†], Carrie A. Kiel [2,†], Corine M. Astroth [3], Kyle G. Dexter [4,5], Frances M. Chase [6] and Erin A. Tripp [7,8]

1   Royal Botanic Gardens, Kew, Richmond, Surrey TW9 3AE, UK
2   Rancho Santa Ana Botanic Garden, Claremont Graduate University, 1500 North College Avenue, Claremont, CA 91711, USA; ckiel@rsabg.org
3   Scripps College, 1030 Columbia Avenue, Claremont, CA 91711, USA; CAstroth7161@scrippscollege.edu
4   School of GeoSciences, University of Edinburgh, Edinburgh EH9 3JN, UK; kyle.dexter@ed.ac.uk
5   Royal Botanic Garden Edinburgh, Edinburgh EH3 5LR, UK
6   National Herbarium of Namibia, Ministry of Environment, Forestry and Tourism, National Botanical Research Institute, Private Bag 13306, Windhoek 10005, Namibia; francesc.nbri@gmail.com
7   Department of Ecology and Evolutionary Biology, University of Colorado, UCB 334, Boulder, CO 80309, USA; erin.tripp@colorado.edu
8   Museum of Natural History, University of Colorado, UCB 350, Boulder, CO 80309, USA
*   Correspondence: i.darbyshire@kew.org; Tel.: +44-(0)20-8332-5407
†   These authors contributed equally.

Received: 1 May 2020; Accepted: 5 June 2020; Published: 11 June 2020

Abstract: *Monechma* Hochst. s.l. (Acanthaceae) is a diverse and ecologically important plant group in sub-Saharan Africa, well represented in the fire-prone savanna biome and with a striking radiation into the non-fire-prone succulent biome in the Namib Desert. We used RADseq to reconstruct evolutionary relationships within *Monechma* s.l. and found it to be non-monophyletic and composed of two distinct clades: Group I comprises eight species resolved within the *Harnieria* clade, whilst Group II comprises 35 species related to the Diclipterinae clade. Our analyses suggest the common ancestors of both clades of *Monechma* occupied savannas, but both of these radiations (~13 mya crown ages) pre-date the currently accepted origin of the savanna biome in Africa, 5–10 mya. Diversification in the succulent biome of the Namib Desert is dated as beginning only ~1.9 mya. Inflorescence and seed morphology are found to distinguish Groups I and II and related taxa in the Justicioid lineage. *Monechma* Group II is morphologically diverse, with variation in some traits related to ecological diversification including plant habit. The present work enables future research on these important lineages and provides evidence towards understanding the biogeographical history of continental Africa.

Keywords: Africa; biome; RADseq; *Monechma*; *Justicia*; phylogeny; plant diversity

---

1. Introduction

The Acanthaceae Juss. (Lamiales) are amongst the most diverse and ecologically important vascular plant families in sub-Saharan Africa. They are, for example, the sixth most species-rich family in the Flora of Ethiopia and Eritrea region, the Flora of Tropical East Africa region (Kenya, Tanzania, Uganda), Mozambique and Namibia; the seventh richest in Cameroon and South Sudan; and the ninth richest in Guinea [1–4]. Lineages of Acanthaceae have diversified in a wide range of habitats ranging from hyper-arid desert to tropical rainforest, and are species-poor only in low-nutrient environments such as on the deep Kalahari Sands of southern Africa and the fynbos of the Cape Floristic Region. In many parts of the continent, Acanthaceae form a dominant constituent of the ground flora such that they provide important ecosystem services and are of economic importance as fodder for livestock and native herbivores [5–7]. Many species of Acanthaceae in sub-Saharan Africa are highly range-restricted and of high conservation concern [6,8–10]. However, despite their obvious importance, our understanding of the diversity and evolutionary history of Acanthaceae is incomplete and many major taxonomic challenges persist [11–16].

One of the most diverse and frequently encountered groups of Acanthaceae in sub-Saharan Africa is the pantropical genus *Justicia* L., taken in a broad sense (i.e., *Justicia* s.l.) [17,18]. Although displaying a large range of morphological diversity, plants of *Justicia* s.l. are readily recognised by the combination of a bilabiate corolla with a rugula (i.e., a stylar furrow on the internal corolla surface), an androecium of two fertile stamens, no staminodes, complex anthers, often with markedly offset thecae and/or with appendages, and 2–4 (–6) colporate pollen with pseudocolpi or with rows of insulae adjacent to the apertures [13,14]. However, recent molecular phylogenetic studies on *Justicia* and allied genera—together comprising the Justicioid lineage—using evidence from six molecular markers [13,14] have demonstrated that *Justicia* s.l. is grossly paraphyletic, with several major, morphologically distinct lineages embedded within it. In order to maintain a broadly circumscribed *Justicia* including morphologically similar taxa such as *Anisotes* Nees, *Anisostachya* Nees, *Monechma* Hochst and *Rungia* Nees, the entire Justicioid lineage would potentially have to be treated as a single genus [13]. This is highly undesirable as it would require subsuming several species-rich genera that are easily separated morphologically, including *Dicliptera* Juss. and *Hypoestes* R. Br. The only plausible alternative, therefore, is to subdivide *Justicia* s.l. into a number of segregate genera [15]. However, only 12–15% of all members of the Justicioid lineage have been phylogenetically sampled to date and many sampling deficiencies need to be addressed before fully informed taxonomic decisions can be made [13].

One such group highlighted as ripe for further taxonomic work is the genus *Monechma* Hochst. s.l. (Figure 1) [13]. *Monechma*, or *Justicia* sect. *Monechma* (Hochst.) T. Anderson, is a group of over 40 species confined to continental Africa and Arabia, with the exception of one species (i.e., the type species, *M. bracteatum* Hochst.) that extends to India (Figure 2). Species of *Monechma* combine the characters of *Justicia* listed above with 2- (rarely 4-) seeded capsules bearing compressed seeds with smooth surfaces [19–21]. However, Kiel et al. [13], upon sampling six species (seven accessions) of *Monechma*, found that this group is not monophyletic and instead separated into two distinct and widely separated clades. *Monechma* Group I, which includes the type species, falls within the Core *Harnieria* clade together with members of *Justicia* sect. *Harnieria* (Solms-Laub.) Benth. (Figure S1). *Monechma* Group II, for which only two species were sampled [13], falls within the Diclipterinae clade, sister to core Diclipterinae: *Kenyacanthus ndorensis* (Schweinf.) I. Darbysh and C.A. Kiel + (*Hypoestes* + *Dicliptera*) (Figure S1).

Attempts to reconcile this unexpected result with morphological evidence [13] suggested that, based on the limited sampling, the two clades could potentially be separated by differences in inflorescence form. *Monechma* Group I was considered to be a predominantly tropical African clade in which the flowers are arranged in 1–few-flowered cymes aggregated into axillary and/or terminal spikes or fascicles, with the bracts markedly differentiated from the leaves. Group II, considered to be a predominantly southern African clade, includes species that have single- or rarely 2-flowered (sub) sessile axillary inflorescences, which can together sometimes form weakly defined terminal spikes, but with the bracts largely undifferentiated from the leaves. In a subsequent study of *Justicia* sect. *Monechma* in Angola [22], this subdivision was expanded upon and the differences in inflorescence form were used to place the majority of Angolan species within Group I. This included both annual, ruderal species, *M. bracteatum* and *M. monechmoides* (S. Moore) Hutch., as well as perennial species of usually fire-prone habitats, such as *M. scabridum* (S. Moore) C.B. Clarke and allies. That study treated these species within *Justicia* in view of the uncertainty over application of the name *Monechma* but noted that species of Group I may ultimately revert to being referred to under *Monechma* following more comprehensive molecular studies [22].

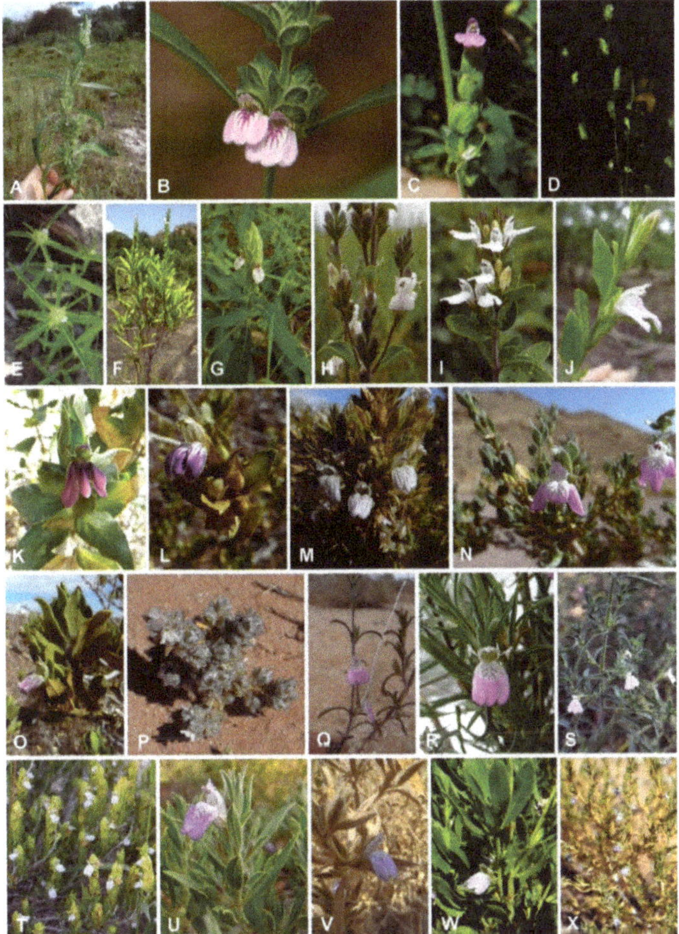

Figure 1. Morphological diversity in *Monechma* s.l. All species pictured have been sampled in the current study. (**A**) *M. monechmoides* (E.A. Tripp, Namibia, collected as *Tripp and Dexter 787*); (**B**) *M. bracteatum* (Mozambique, B. Wursten); (**C**) *M. debile* (C.A. Kiel, Kenya, *Kiel 173*); (**D**) *Justicia* sp. B of Flora Zambesiaca (B. Wursten, Mozambique, *Wursten 1792*); (**E**) *M. ciliatum* (K. Schumann, Burkina Faso); (**F**) *M. ndellense* (A. Thiombiano, Burkina Faso); (**G**) *M. depauperatum* (W. McCleland, Mali); (**H**) *M. rigidum* (D.J. Goyder, Angola, *Goyder 8210*); (**I**) *M. virgultorum* (D.J. Goyder, Angola, *Goyder 8471*); (**J**) *M. serotinum* (E.A. Tripp, Namibia, *Tripp et al. 4068*); (**K**) *M. grandiflorum* (E.A. Tripp, Namibia *Tripp et al. 2034*); (**L**) *M. calcaratum* (E.A. Tripp, Namibia, *Tripp et al. 2043*); (**M**) *M. distichotrichum* (E.A. Tripp, Namibia, *Tripp et al. 2072*); (**N**) *M. leucoderme* (E.A. Tripp, Namibia, *Tripp et al. 2083*); (**O**) *M. mollissimum* (E.A. Tripp, Namibia, *Tripp et al. 2071*); (**P**) *M. desertorum* (L. Nanyeni, Namibia); (**Q–S**) *M. divaricatum*; (**Q**) (E.A. Tripp, Namibia, *Tripp and Dexter 885*); (**R**) (E.A. Tripp, Namibia, *Tripp and Dexter 779*); (**S**) (I. Darbyshire, Namibia); (**T**) *M. genistifolium* (I. Darbyshire, Namibia); (**U & V**) *M. cleomoides*; (**U**) (E.A. Tripp, Namibia, *Tripp and Dexter 829*); (**V**) (E.A. Tripp, Namibia, *Tripp et al. 1960*); (**W**) *M. tonsum* (E.A. Tripp, Namibia, *Tripp and Dexter 813*); (**X**) *M. salsola* (I. Darbyshire, Namibia, *Klaassen et al. 2537*). E, F & G reproduced from S. Dressler, M. Schmidt and G. Zizka, African Plants—A Photo Guide: www.africanplants.senckenberg.de [23] with kind permission from the authors and photographers.

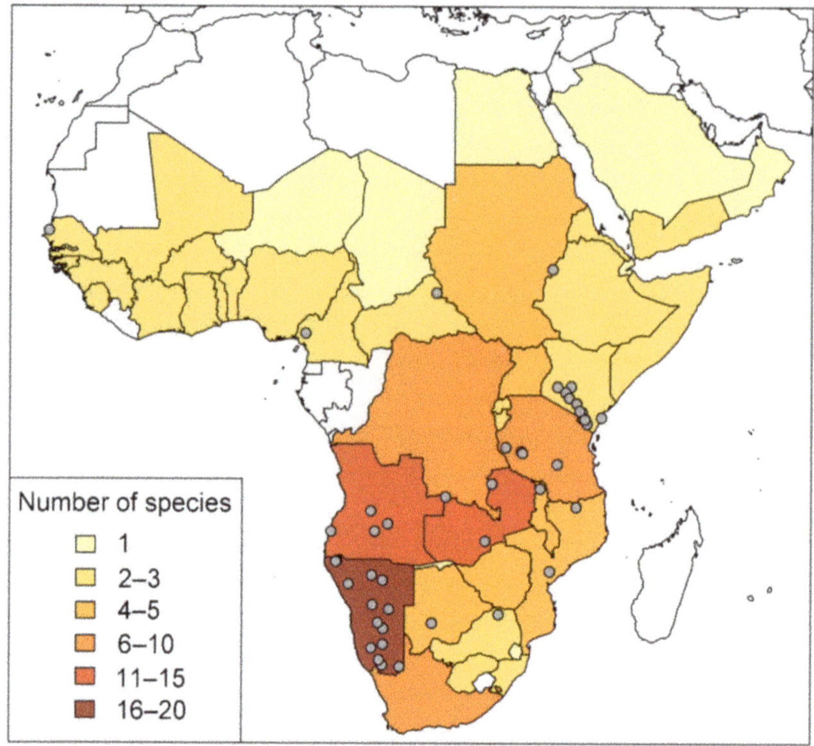

Figure 2. Distribution of *Monechma* s.l. in Africa and Arabia; species richness per TDWG Level 3 geographic region. Note: this is the global distribution of *Monechma* s.l. except that one species (*M. bracteatum*) extends to India. Grey dots represent the samples used in the current study (see Table 1).

Table 1. Taxon and source of plant material from which DNA was extracted for sequencing. Taxa are listed in alphabetical order by genus and species.

Taxon	Source Specimen	Country	Latitude	Longitude
Dicliptera maculata Nees subsp. *usambarica* (Lindau) I. Darbysh.	*Kiel et al. 157* (RSA)	Kenya	−0.1791	35.6317
Dicliptera paniculata (Forssk.) I. Darbysh.	*Kiel et al. 166* (RSA)	Kenya	−2.6910	38.1639
Hypoestes forskaolii (Vahl) R. Br.	*Kiel et al. 144* (RSA)	Kenya	−1.8087	37.5864
Hypoestes triflora (Forssk.) Roem. & Schult.	*Kiel et al. 151* (RSA)	Kenya	−0.7033	36.4346
Justicia anagalloides (Nees) T. Anderson	*Kiel et al. 174* (RSA)	Kenya	−3.4144	38.4262
Justicia attenuifolia Vollesen	*Golding et al. 8* (K)	Mozambique	−12.1739	37.5494
Justicia cordata (Nees) T. Anderson	*Kiel et al. 159* (RSA)	Kenya	−2.5514	37.8933
Justicia cubangensis I. Darbysh. & Goyder	*Goyder et al. 8068* (K)	Angola	−14.5897	16.9072
Justicia eminii Lindau	*Bidgood et al. 930* (K)	Tanzania	−7.9167	35.6000
Justicia fanshawei Vollesen	*Smith et al. 2010* (K)	Zambia	−9.8529	28.9441
Justicia flava (Forssk.) Vahl	*Kiel et al. 146* (RSA)	Kenya	−1.8082	37.5765
Justicia heterocarpa T. Anderson	*Kiel et al. 158* (RSA)	Kenya	−1.2745	36.8146
Justicia kirkiana T. Anderson	*Kiel et al. 177* (RSA)	Kenya	−3.8407	38.6681
Justicia odora (Forssk.) Lam.	*Tripp et al. 4073* (COLO)	Namibia	−17.6041	12.8872
Justicia phyllostachys C.B. Clarke	*Bidgood et al. 6871* (K)	Tanzania	−6.7833	32.0667
Justicia platysepala (S. Moore) P.G. Mey.	*Tripp and Dexter 4119* (COLO)	Namibia	−22.3833	18.4073
Justicia platysepala (S. Moore) P.G. Mey.	*Tripp et al. 6907* (COLO)	Angola	−12.8929	13.4947
Justicia platysepala (S. Moore) P.G. Mey.	*Tripp et al. 6919* (COLO)	Angola	−14.9700	12.9040
Justicia pseudorungia Lindau	*Kiel et al. 185* (RSA)	Kenya	−3.2222	40.1218
Justicia sp. B. of Flora Zambesiaca	*Bester 11112* (K)	Mozambique	−18.5622	34.8731

Table 1. *Cont.*

Taxon	Source Specimen	Country	Latitude	Longitude
Justicia striata (Klotzsch) Bullock	*Kiel et al. 145* (RSA)	Kenya	−1.8082	37.5765
Justicia tetrasperma Hedrén	*Kahurananga et al. 2582* (K)	Tanzania	−6.1994	30.3536
Justicia tricostata Vollesen	*Bidgood et al. 5606* (K)	Tanzania	−8.4500	31.4833
Justicia tricostata Vollesen	*Gillis 11441* (RSA)	Zambia	−15.5470	28.2472
Justicia unyorensis S. Moore	*Kiel et al. 163* (RSA)	Kenya	−2.5514	37.8933
Justicia vagabunda Benoist	*Tripp et al. 1544* (RSA)	China	21.9449	101.2735
Kenyacanthus ndorensis (Schweinf.) I. Darbysh. & C.A. Kiel	*Luke et al. 17084* (K)	Kenya	−0.1499	37.0238
Monechma australe P.G. Mey.	*Tripp et al. 2028* (RSA)	Namibia	−23.7117	17.2600
Monechma bracteatum Hochst.	*Kiel et al. 161* (RSA)	Kenya	−2.5514	37.8933
Monechma bracteatum Hochst.	*Friis et al. 13545* (K)	Ethiopia	11.5285	35.1075
Monechma calcaratum Hochst.	*Tripp and Dexter 2043* (RSA)	Namibia	−25.8755	17.7929
Monechma ciliatum Hochst. ex Nees	*Merklinger 2013-9-55* (K)	Senegal	15.3181	−16.7758
Monechma cleomoides C.B. Clarke	*Klaassen et al. 2530* (K)	Namibia	−21.2978	15.2803
Monechma cleomoides C.B. Clarke	*Tripp et al. 1995* (RSA)	Namibia	−17.8023	12.3261
Monechma cleomoides C.B. Clarke	*Tripp et al. 1960* (RSA)	Namibia	−19.8212	14.1870
Monechma cleomoides C.B. Clarke	*Tripp et al. 1999* (RSA)	Namibia	−17.5193	12.2674
Monechma debile Nees	*Friis et al. 10459* (K)	Ethiopia	13.8167	39.5500
Monechma debile Nees	*Kiel et al. 173* (RSA)	Kenya	−3.3496	38.4483
Monechma depauperatum C.B. Clarke	*Etuge 4446r* (K)	Cameroon	5.0833	9.7167
Monechma desertorum C.B. Clarke	*Oliver et al. 6379* (K)	Namibia	−27.4028	17.3833
Monechma distichotrichum P.G. Mey.	*Tripp et al. 2067* (RSA)	Namibia	−28.0878	19.5131
Monechma distichotrichum P.G. Mey.	*Tripp et al. 2072* (RSA)	Namibia	−27.9074	17.6788
Monechma divaricatum C.B. Clarke	*Tripp and Dexter 808* (RSA)	Namibia	−18.7071	17.2921
Monechma divaricatum C.B. Clarke	*Tripp and Dexter 783* (RSA)	Namibia	−19.5546	17.7329
Monechma divaricatum C.B. Clarke	*McDade et al. 1275* (RSA)	South Africa	−22.8833	29.6667
Monechma divaricatum C.B. Clarke	*Tripp et al. 1970* (RSA)	Namibia	−19.6156	13.2550
Monechma divaricatum C.B. Clarke	*Tripp et al. 2023* (RSA)	Namibia	−23.3475	17.0788
Monechma divaricatum C.B. Clarke	*Tripp et al. 1961* (RSA)	Namibia	−19.8429	14.1279
Monechma divaricatum C.B. Clarke	*Tripp et al. 2029* (RSA)	Namibia	−23.7117	17.2600
Monechma divaricatum C.B. Clarke	*Tripp et al. 2039* (RSA)	Namibia	−26.4395	18.1855
Monechma divaricatum C.B. Clarke	*Tripp and Dexter 4800* (COLO)	Namibia	−20.3351	17.5604
Monechma genistifolium C.B. Clarke	*Tripp and Dexter 775* (RSA)	Namibia	−21.9340	16.6867
Monechma genistifolium C.B. Clarke	*Wanntorp & Wanntorp 339* (K)	Namibia	−21.5125	16.0314
Monechma grandiflorum Schinz	*Tripp and Dexter 2034* (RSA)	Namibia	−24.3024	17.8223
Monechma incanum C.B. Clarke	*Mott 1124* (K)	Botswana	−23.7656	22.8097
Monechma incanum C.B. Clarke	*Puff 780416-2/2* (RSA)	South Africa	−27.9471	22.6925
Monechma leucoderme C.B. Clarke	*Tripp and Dexter 2044* (RSA)	Namibia	−25.8755	17.7929
Monechma leucoderme C.B. Clarke	*Tripp et al. 2083* (RSA)	Namibia	−26.2326	16.5967
Monechma mollissimum (Nees) P.G. Mey.	*Balkwill et al. 11787* (RSA)	South Africa	−28.9489	18.2433
Monechma mollissimum (Nees) P.G. Mey.	*Tripp et al. 2071* (RSA)	Namibia	−27.9231	17.7338
Monechma monechmoides (S. Moore) Hutch.	*Aiyambo et al. 323* (K)	Namibia	−19.4713	17.7469
Monechma monechmoides (S. Moore) Hutch.	*Tripp and Dexter 785* (RSA)	Namibia	−19.4713	17.7469
Monechma monechmoides (S. Moore) Hutch.	*Dingham 11019* (K)	Zambia	15.1667	27.1667
Monechma ndellense (Lindau) J. Miège & Heine	*Harris & Fay 2150* (K)	C.A.R.	9.1667	23.2167
Monechma rigidum S. Moore	*Goyder 8210* (K)	Angola	−12.5683	16.4931
Monechma salsola C.B. Clarke	*Klaassen et al. 2537* (K)	Namibia	−19.2528	14.0044
Monechma salsola C.B. Clarke	*Klaassen et al. 2544* (K)	Namibia	−19.1944	13.0861
Monechma salsola C.B. Clarke	*Tripp and Dexter 6934* (COLO)	Angola	−14.5999	12.3703
Monechma scabridum S. Moore	*Congdon 584* (K)	Zambia	−11.1664	24.1850
Monechma serotinum P.G. Mey.	*Tripp et al. 4066* (COLO)	Namibia	−17.5117	12.9696
Monechma spartioides (T. Anderson) C.B. Clarke	*Tripp et al. 2064* (RSA)	Namibia	−28.0878	19.5131
Monechma sp.	*Tripp and Dexter 834* (RSA)	Namibia	−17.6070	12.9523
Monechma subsessile C.B. Clarke	*Bidgood et al. 6793* (K)	Tanzania	−6.6167	31.9333
Monechma tonsum P.G. Mey.	*Nyatoro et al. 29* (K)	Namibia	−18.1367	13.8953
Monechma tonsum P.G. Mey.	*Tripp and Dexter 813* (RSA)	Namibia	−18.9546	16.6243
Monechma varians C.B. Clarke	*Synge WC437* (K)	Malawi	−10.3500	33.8833
Monechma virgultorum S. Moore	*Goyder 8471* (K)	Angola	−13.8519	18.2589

1.1. Ecological Importance of Members of Monechma s.l.

Members of *Monechma* s.l. are widespread in sub-Saharan Africa (Figure 2). While a significant number of species of the group occur in fire-prone vegetation corresponding to the savanna biome, the group becomes particularly abundant and diverse (18+ species) in the deserts and shrublands of southwest Africa, centered in southern Angola, Namibia and the Northern Cape region of South

Africa [24], which represents one of the main extensions of the succulent biome in Africa [25] (Figure 3). The tropical succulent biome is less well-known than the tropical savanna biome; both experience seasonality in water availability, but the succulent biome differs from savanna in rarely experiencing fire [26]. Within southwest Africa, *Monechma* frequently forms a major component of the dominant ground flora, often in combination with one or more of three other distantly related lineages in the Acanthaceae family that have diversified independently in this region: *Barleria* L. (Barlerieae) [7], *Blepharis* Juss. (Acantheae) [5] and *Petalidium* Nees (Ruellieae) [6], with both *Barleria* and *Petalidium* represented by over 25 spp. in Namibia alone [27]. The parallel radiation of species in these four genera within the succulent biome in southwest Africa is remarkable, and together result in the Acanthaceae being amongst the most important plant families in the region. In view of the exceptional ecological importance of these genera, it is essential that we have a strong understanding of the species diversity and evolutionary history of these groups. Taxonomic studies of the Namibian radiation of *Monechma* are ongoing as part of the *Flora of Namibia* programme [28]; however, phylogenetic investigation of the evolutionary history of the group has been lacking to date.

Figure 3. The habitat and abundance of *Monechma* in Namibia. (**A**) *M. genistifolium* (bright green) together with *Petalidium englerianum* (Schinz) C.B. Clarke (silver-green) near Outjo (I. Darbyshire); (**B**) *M. tonsum* together with *Petalidium variabile* C.B. Clarke s.l. near Opuwo, collected as *Nyatoro et al. 29* (I. Darbyshire); (**C**) *M. spartioides* c. 30 km W of Ariamsvlei (E.A. Tripp, collected as *Tripp et al. 2064*); (**D**) *M. salsola* near Umbaadjie, collected as *Klaassen et al. 2537* (I. Darbyshire).

Elsewhere in tropical Africa, members of *Monechma* s.l. can be an important constituent of the fire-prone savanna biome of both the Sudanian and Zambesian phytogeographic regions [29], for example *M. depauperatum* (T. Anderson) C.B. Clarke in the Sudanian region and *M. scabridum* in the Zambesian region. Other species such as *M. bracteatum* and *M. monechmoides* favour open habitats with high light availability and so can be common in disturbed, ruderal environments.

1.2. Aims of the Present Study

The present study intends to reconstruct evolutionary relationships within *Monechma* s.l. in the context of the wider classification of the Justicioid lineage and towards understanding the diversification of this ecologically important lineage. A RADseq phylogenetic approach is used in light of the considerable success that this method has provided in resolving phylogenetic relationships within other major lineages of Acanthaceae, including *Petalidium* [6], *Louteridium* S. Watson [30], Ruellieae [31], *Barleria* [32] and New World *Justicia* [33]. The sampling of species of *Monechma* s.l. is here expanded to include ca. 75% of the accepted taxonomic diversity and, in many cases, to include multiple accessions per species with the goal of assessing reciprocal monophyly of such lineages. Specifically, we aim to (a) test prior delimitation of the two clades of *Monechma*; (b) identify and/or confirm morphological traits that diagnose the recognised clades; (c) present a first assessment of the biogeographical history of the genus; (d) place all known species of *Monechma* s.l. into a taxonomic context through a combination of molecular and morphological evidence; and (e) provide a phlyogenetic framework to assist with ongoing and future monographic and floristic work on *Monechma* s.l. and allies in the Justicioid lineage.

2. Materials and Methods

2.1. Sampling

In total, 80 accessions were sampled. Of these, 59 accessions represent 32 of the total 42 species (76%) currently accepted in *Monechma* or in *Justicia* sect. *Monechma*, plus three taxa that are unidentified to species or represent currently undescribed species. The sampling was designed to capture the full range of morphological variation within *Monechma* s.l. as well as to include two or more accessions of morphologically variable species wherever possible. To help delimit broader-scale relationships, we also included 29 accessions spanning major clades of the Justicioid lineage [13]. *Justicia pseudorungia* Lindau of the *Rungia* clade [13] was used as an outgroup for rooting our phylogenetic hypothesis. Leaf tissue for molecular analyses was sampled from either field-collected plant material dried in silica gel or herbarium specimens. Table 1 includes taxon names, source locality and voucher number for all accessions used in this study excluding the removed samples (see Section 2.3); these are mapped on Figure 2.

2.2. DNA Isolation and Sequencing Methods

ddRADseq data (double digest restriction-associated DNA) were used to reconstruct phylogenetic relationships among *Monechma*. At the University of Colorado (Boulder, CO, USA) and Rancho Santa Ana Botanic Garden (RSABG) (Claremont, CA, USA), DNA was extracted from dried leaf tissue using a CTAB protocol [34]. ddRAD libraries were constructed at RSABG using a modified version of that used in [6], which was originally adapted from [35]. A full description of this protocol is published in [6], with details briefly outlined here. All genomic DNA was normalized to ~30 ng/μL before digestion and library construction. Extracted DNA underwent double restriction enzyme digestion using *EcoRI* and *MseI* for 3 h at 37 °C followed by 65 °C for 45 min. Illumina sequencing oligos together with in-line, variable-length barcodes were annealed to the *EcoRI* cut site and ligated onto digested fragments. Illumina oligos were similarly annealed to the *MseI* cutsite. Barcoded ligation products were pooled and cleaned using a Qiagen gel extraction kit. We excised fragments from the gel between 200–700 bp to reduce the effects of dimer and to provide more precise amplification of the targeted region. The gel-purified ligations were amplified using the following PCR reaction: 8.6 μL of water, 4 μL of Phusion HF buffer, 0.5 μL of each Illumina primer (10 μM), 0.6 μL DMSO, 0.6 μL DNTPs, 0.2 μL Phusion. Fifteen cycles of PCR were conducted to amplify the cleaned, ligated products. The reaction was repeated once to ameliorate stochastic differences in PCR amplification. Agarose gels were used to assess amplification and size of the PCR products and amplicon concentrations were evaluated using a Qubit fluorometer 2.0. The custom-tagged products of the PCR reactions were pooled and sent to the University of Colorado's Biofrontiers Next-Gen Sequencing Facility for quality control and further size selection. BluePippin was used to select a fragment range between 200 and 500 bp to reduce the sequenced genome. Libraries from the 80 samples were pooled to yield a final combined library that was submitted for 1 × 75 sequencing on an Illumina NextSeq v2 High Output Sequencer at Biofrontiers.

2.3. Phylogenetic Reconstruction

We assessed sequencing quality of raw data using FastQC [36]. Data were filtered, trimmed, and demultiplexed using iPYRAD 0.9.31 [37,38]. Of the 80 taxa sampled, four accessions—*Monechma* sp. (specimen: *Tripp et al. 1985*), *Rhinacanthus angulicaulis* I. Darbysh. (*Kiel et al. 170*), *Justicia flava* (Forssk.) Vahl (*Kiel et al. 146*) and *Justicia striolata* Mildbr. (*Congdon et al. 794*)—were removed because of too few loci (i.e., values < 40). Information on the number of ddRAD reads per sample and loci in the assembly for each accession sampled in our study are provided in Table S1. As a result, our final sampling contained 76 accessions, which included 58 accessions of *Monechma* representing 34 taxa (32 accepted species). Of these taxa, 13 were represented by two or more accessions to account for species with broad geographical distributions and/or variation in morphology (Table 1). The de novo assembly parameters for our final dataset are as follows: the minimum required sequence length (to retain a read) = 35 bp; minimum coverage for retaining a cluster = 6; maximum low quality bases = 5; clustering threshold (level of sequence similarity in which two sequences are identified as homologous) = 0.90; minimum number of samples that must have data at a given locus to be retained = 20; maximum number of alleles per site in consensus sequence = 2. We also conducted 3 additional de novo assemblies exploring the number of minimum samples required to retain a locus (i.e., 4, 10, 30). The final RADseq phylogenomic dataset is available in Sequence Read Archive (SRA) under the BioProject number PRJNA635173.

2.4. Phylogenetic Analyses

We implemented two approaches for estimating phylogenetic relationships among *Monechma* s.l.: (1) a Maximum Likelihood (ML) analysis using the concatenated RAD sequence data from all loci derived from the iPYRAD [37,38] assembly and (2) a coalescent-based approach using quartet-based phylogenetic inference under a multispecies coalescent theory framework that used the concatenated RAD sequence data described above, but randomly sampled one SNP per locus. We conducted our ML analyses using IQ-TREE 1.6.10 [39]. The best model of nucleotide substitution and across-site heterogeneity in evolutionary rates was inferred using ModelTest-NG 0.1.5 [40]. The best-fit model was selected based on the corrected Akaike's information criterion. Node and branch supports were obtained from 1000 nonparametric bootstrap replicates under the best inferred model (GTR + G). We constructed quartet-based coalescent phylogenetic inferences using the program Tetrad [41] in iPYRAD [37,38] and assessed node support with 1000 bootstraps. The SVDquartets algorithm [42], implemented in Tetrad [41], uses multi-locus unlinked SNP data to infer the topology among all possible subsets of four samples under a coalescent model. The resulting set of quartet trees are combined and constructed into a species tree. Because the underlying model assumes that the examined SNPs are unlinked, Tetrad subsamples a single SNP from every locus separately for every quartet set in the analysis from the .snps.hdf5 file produced from the iPYRAD output and repeats this subsampling method independently in each bootstrap replicate. This method maximizes the number of unlinked SNP information in the analysis. For both ML and Tetrad analyses, we considered branches to be supported when bootstrap values were >90%, while bootstrap values < 70% were considered unsupported.

2.5. Hypothesis Testing

Six alternative phylogenetic hypotheses were examined using the Shimodaira Approximately Unbiased (AU) tests [43]. Constraint trees were constructed in Mesquite v.2.72 [44]. For each constraint, all aspects of relationships were constructed as a single polytomy, with the exception of the hypothesis under consideration. The constraint trees were loaded into IQ-TREE [39] and run with the settings and model as described above. The best trees from the unconstrained and constrained analyses were combined into a single file and loaded into IQ-TREE and likelihood scores were compared using the AU test with RELL-optimization and 10,000 bootstrap replicates.

2.6. Divergence Time Estimation

To provide temporal context to the evolutionary history of *Monechma* and close relatives, we estimated divergence times using the most likely tree from our concatenated ML analysis. We pruned this tree to contain a single representative for each ingroup taxon, resulting in a total of 49 species. The singleton tree was rate-smoothed and ultrametricized using penalized likelihood under a relaxed model, where rates are uncorrelated across branches [45] as implemented with the *chronos* function in ape v 5.1 [46] of R v 3.6.0 ("Planting of a Tree") [47]. A best-fit smoothing parameter (lambda) of 1.0 was selected following the cross-validation approach and chi-square test as implemented in treePL [48], testing eight values between 0–1000 distributed on a log-scale. A single fossil calibration for a minimum age date of 11.5 my was used to constrain the most recent common ancestor of the Justicioid lineage. This fossil was previously assessed as both reliably identified and dated [49]. Fossil #32 [49] from the Middle Miocene is a dicolporate pollen grain with distinctive round insulae that laterally flank the apertures [50]; the latter of these traits is known only among Justicioids [13,14,51]. We also used a 35 my maximum date for our calibration, which is the estimated age for Justicieae as a whole [49].

2.7. Biome Evolution and Climatic Niche

We reconstructed an ancestral biome state of lineages to elucidate the history of biome occupancy and biome switching in *Monechma* s.l. For all taxa in our ultrametric tree, species presence/absence in four biomes was determined based on ecoregions [52], as follows: (1) tropical and subtropical grasslands, savannas and shrublands (hereafter savanna biome); (2) deserts and xeric shrublands (hereafter succulent biome); (3) tropical and subtropical broadleaf forests (hereafter forest biome), and (4) montane grasslands and shrublands (hereafter montane biomes). Ancestral state reconstructions were implemented with the *rayDISC* function in the corHMM 1.13 package [53]. This function assumes a constant rate of evolution across all branches and permits polymorphic character states that account for the probability of either state when calculating the likelihood at ancestral nodes. We compared two distinct Markov models of discrete character evolution: the equal rates (ER) or Mk model, which assumes a single rate of transition among all possible states, and the all rates different (ARD) or the AsymmMk model [54,55], which allows different rates for each possible transition. We also examined a symmetrical model (SYM), which specifies equal rate transitions in either direction between pairs of states but permits different rates between different pairs. Model fit was tested by comparing AICc values, from which we selected the model that best fits the data while minimizing the number of parameters [56].

Given asymmetrical patterns of standing diversity in *Monechma* s.l., specifically far greater species richness and abundance in southwestern portions of the range of this lineage, we sought to delimit climatic niche preferences among species throughout the range. We first downloaded 19 WorldClim Bioclimatic variables available in the WorldClim database [57] at 30 arc-seconds resolution [58]. We then extracted bioclimatic data for taxa in our ultrametric tree using latitude and longitude of collections in the R package raster [59]. We visualized changes in two climatic variables: BIO7 = temperature annual range (BIO5 − BIO6: minimum temperature of the warmest month − minimum temperature of the coldest) and BIO12 = annual precipitation (mm), using the *contmap* function in the package phytools [60]. The mapping is accomplished by estimating ancestral states at internal nodes using ML with the *fastAnc* function and then interpolating the states along each edge using Equation (2) of [61]; see [62].

2.8. Morphological Studies

A survey of morphological traits that have been found to be taxonomically informative in past studies of both *Monechma* s.l. and the wider Justicioid lineage was conducted for all relevant taxa in order to interpret results of the RADseq analyses. We focused on the following morphological traits: plant habit, inflorescence form, details of the androecium including arrangement of anther thecae and details of the staminal appendages, pollen morphology, and seed number, size, shape and indumentum. Most observations were made on herbarium specimens held at K, RSA and COLO (herbarium abbreviations follow [63]) but with additional observations made via access to digital images of type specimens on JSTOR Global Plants [64] and other online repositories of herbarium specimen images. For pollen morphology, unacetolyzed pollen from selected taxa was mounted on aluminum stubs using double-sided sticky tape and coated with gold using a PELCO SC-7 system (Ted Pella, Redding, CA, USA). The coated samples were observed at 10 kV on a Hitachi SU3500 (Hitachi, Tokyo, Japan) scanning electron microscope (SEM) at Rancho Santa Ana Botanic Garden. Chromosome number was also considered through reference to relevant cytological studies.

The geographic distribution of each accepted taxon was delimited using the Level 3 codes of the TDWG geographic scheme for recording plant distributions [65].

3. Results

3.1. Phylogenetic Results

The phylogenies inferred using ML for each of the four concatenated data sets were congruent despite variation in the proportion of missing data (Figure 4, Figures S2–S4). The datasets containing more missing data (i.e., larger alignment files with lower min tax values) yielded similar or identical topologies to the datasets containing fewer missing data (i.e., smaller alignment files with higher min tax values; Figures S2 and S3). However, topologies of the latter, in particular the dataset with minimum samples per locus = 30, had lower bootstrap supports for relationships along the backbone of the phylogeny (Figure S4). We here present the results of the concatenated dataset with the minimum samples per locus set at 20 (Figure 4), which contained 5718 loci and 468,892 SNPs. We chose this assembly because it contains the least amount of missing data without losing resolution (see results from Tetrad analysis, below) while also maximizing the amount of genome data utilized. The coalescent analysis (Figure S5) using the final genotype matrix from the de novo assembly (468,892 SNPs and 20,000,000 quartet sets) resulted in a similar species-level topology to that inferred from the concatenated ML analysis of data. However, the resulting topology inferred from the Tetrad analysis exhibits low resolution along the backbone and thus ambiguous relationships among major clades. Overall, there were no strongly supported topological conflicts between the ML vs. Tetrad analyses (Figure 4 and Figure S5).

Overall, the phylogenetic results from all analyses concur with the findings of the earlier studies [13] (Figure S1) that *Monechma* s.l. is polyphyletic and that species previously placed in this genus (or in *Justicia* sect. *Monechma*) are resolved in one of two clades, with the exception of *M. varians* (see below). Our data reject strict monophyly of *Monechma* s.l. ($p < 0.001$; Table 2).

Monechma Group I (ML: 100% BS; Figure 4) was resolved in the *Harnieria* clade of the Justicioid lineage and is here composed of six species (*M. bracteatum*, *M. debile*, *M. monechmoides*, *Justicia eminii*, *J. tetrasperma* and *J. sp. B* of Flora Zambesiaca; [18]). This clade is sister to *Justicia odora* of *Justicia* sect. *Harnieria*, and these together are sister to the other four sampled members of sect. *Harnieria*. Results from an AU test do not reject the monophyly of *Justicia* sect. *Harnieria* (i.e., *J. unyorensis*, *J. heterocarpa*, *J. striata*, *J. phyllostachys* and *J. odora*; $p = 0.393$). Species of *Monechma* Group I for which two or more accessions were sampled (i.e., *M. debile*, *M. bracteatum*, *M. monechmoides*) were each resolved as reciprocally monophyletic (Figure 4).

Monechma Group II (ML: 98% BS; Figure 4) was resolved as a sister to all sampled members of the Diclipterinae clade (i.e., two species each of *Dicliptera* and *Hypoestes*; *Justicia vagabunda* and *Kenyacanthus ndorensis*). *Monechma* Group II is here composed of 26 species and includes two major clades. The first clade (ML: 100% BS; Figure 4) consists of six tropical African species (*M. ciliatum*, *M. depauperatum*, *M. ndellense*, *M. scabridum*, *M. subsessile* and *Justicia attenuifolia*). The second major clade (ML: 100% BS; Figure 4) contains the remaining 17 sampled species of tropical and southern African *Monechma*, primarily those of the succulent biome radiation, in addition to *Justicia fanshawei*, *J. cubangensis*, and *J. tricostata*, all of which were described in *Justicia* sect. *Monechma*. Most species of *Monechma* Group II for which two or more accessions were sampled (i.e., *M. distichotrichum*, *M. divaricatum*, *M. genistifolium*, *M. incanum*, *M. leucoderme*, *M. mollissimum* and *M. salsola*) were each resolved as reciprocally monophyletic (Figure 4). Sampled accessions of *Monechma cleomoides* and *M. tonsum* were not resolved as reciprocally monophyletic and instead were resolved as part of a clade containing *M. genistifolium*, *M. australe*, and *M. salsola*. Results of an AU test also reject the monophyly of *M. cleomoides* and *M. tonsum* ($p < 0.001$; Table 2).

Monechma varians was here resolved as sister to *Justicia kirkiana* of the *Tyloglossa* clade (ML: 96% BS; Figure 4). Together, these taxa are sister to Core Diclipterinae + *Monechma* Group II and the *Harnieria* clade, but with weak support (ML: 80% BS; Figure 4).

Justicia platysepala, which was originally placed in *Monechma* (*M. platysepalum* S. Moore) but more recently has been included in *Justicia* [26,66], was here resolved in a clade consisting of *J. anagalloides* and *J. cordata* (ML: 94% BS; Figure 4). This clade is sister to all other sampled in-group taxa in our dataset.

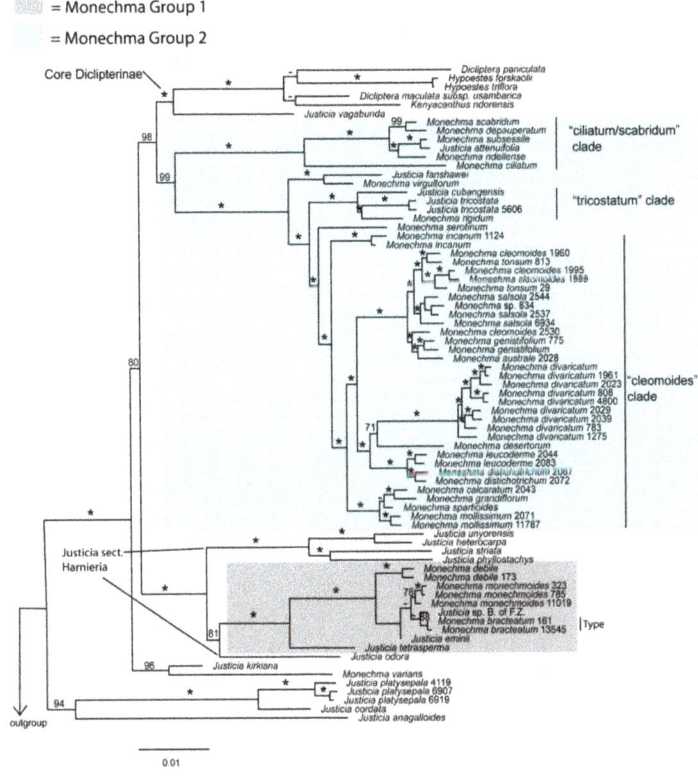

Figure 4. The most likely phylogenetic hypothesis for relationships among *Monechma* s.l. generated from ddRADseq loci. *Monechma* s.l. is not monophyletic and is resolved in two major clades: *Monechma* Group I (grey box) and *Monechma* Group II (blue box). The type species, *Monechma bracteatum*, is denoted in Group I. Collection numbers are listed after species names where multiple accessions were sampled. Asterisks [*] indicate 100% ML bootstrap and dashes [-] indicate <70% ML bootstrap.

Table 2. Results of alternative phylogenetic hypothesis testing using the Shimodaira Approximately Unbiased (AU) test among *Monechma*. H0 = results from present study; H1 = alternative hypotheses based on earlier classification or morphological patterns.

Hypothesis	logL	D (logL)	Reject?	*p*-Value
H0. *Monechma* s.l. is not monophyletic	−1201,860.278			
H1. *Monechma* s.l. (excluding *M. varians*) is monophyletic	−1209,690.731	7830.5	Yes	<0.0001
H0. *Justicia* sect. *Harnieria* is not monophyletic	−1201,860.278			
H1. *Justicia* sect. *Harnieria* is monophyletic, i.e., *Monechma* Group I is not embedded within this section	−1201,862.068	1.7897	No	0.393
H0. *M. ciliatum* is a member of *Monechma* Group II	−1201,860.278			
H1. *M. ciliatum* is not a member of *Monechma* Group II	−1208,510.036	6649.8	Yes	<0.0001
H0. *M. cleomoides* + *M. tonsum* is not monophyletic	−1201,860.278			
H1. *M. cleomoides* including *M. tonsum* is monophyletic	−1203,255.026	1394.7	Yes	<0.0001
H0. *M. cleomoides* is not monophyletic	−1201,860.278			
H1. *M. cleomoides* is monophyletic	−1204,005.563	2145.3	Yes	<0.0001
H0. *M. tonsum* is not monophyletic	−1201,860.278			
H1. *M. tonsum* is monophyletic	−1203,255.026	1394.7	Yes	<0.0001

3.2. Divergence Times

Our divergence time analyses using penalized likelihood estimated that *Monechma* Group I plus *Justicia odora* of the *Harnieria* clade originated around 22 mya (stem group) and began to diversify around 18 mya (crown), with *Monechma* Group I specifically diversifying at approximately 12.3 mya (crown; Figure 5). Our analyses estimate that *Monechma* Group II originated around 22.5 mya (stem) and began diversifying around 13.4 mya. Within Group II, however, the succulent biome radiation is estimated to have begun diversifying as recently as 1.9 mya (Figure 5).

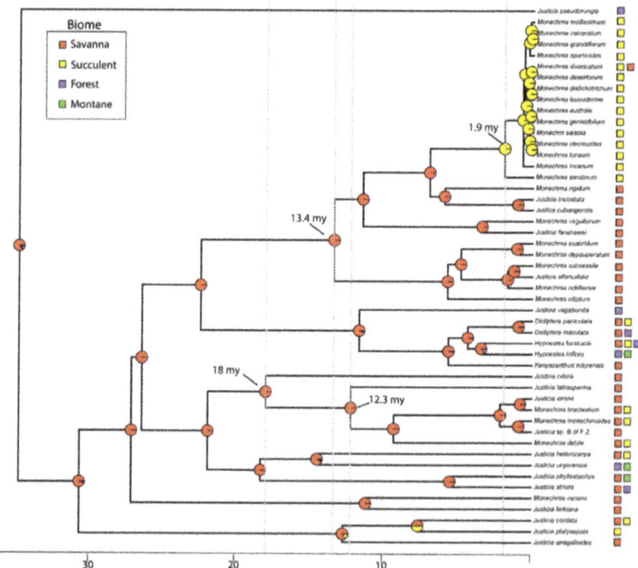

Figure 5. Divergence time estimation from penalized likelihood and biome evolution for *Monechma* Groups I and II. Ancestral state reconstruction: circles at nodes are color coded to reflect ancestral character states with sizes of differently colored wedges indicating likelihood of presence of each state at that node.

3.3. Biome Evolution and Climatic Niche

In our analyses examining transitions among biomes, the common ancestor of the *Harnieria* clade + *Monechma* Group I was most likely distributed in savannas (ER model based on AICc values; Figure 5; Table S2). The ancestor of *Monechma* Group II similarly most likely occupied savannas, with subsequent shifts into the succulent biome including deserts and xeric shrublands (Figure 5). Throughout *Monechma* s.l. and allies, our results suggest there have been rare shifts to tropical forests and montane environments from savanna ancestors (Figure 5).

Ancestral state reconstruction of climatic variables suggests marked shifts in temperature and precipitation regimes during evolution of the succulent biome radiation of Group II. Species in this group (i.e., the *"cleomoides"* clade) have diversified into drier habitats with greater ranges of temperature extremes in comparison to the others in Group II (Figure 6) as well as the remainder of sampled species in our analyses, including those of *Monechma* Group I.

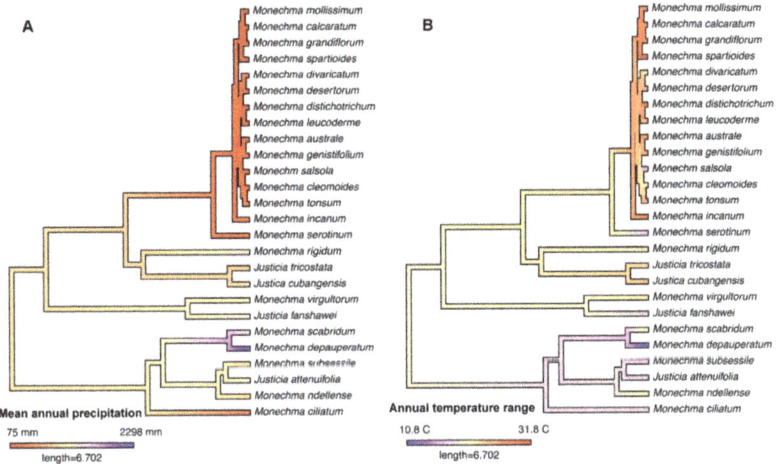

Figure 6. Visualization of WorldClim variables along the nodes and branches of *Monechma* Group II. (**A**) mean annual precipitation (BIO12); (**B**) temperature annual range (BIO7). Species with red color represent species from (**A**) drier climates with a (**B**) greater range of temperature extremes. Raw values were plotted using the *contMap* function in package phytools.

3.4. Taxonomically Informative Morphological Traits

Our analyses indicated inflorescence form and seed morphology are the most informative morphological characters for separation of *Monechma* Groups I and II. Characters that were found not to be diagnostic for these two clades were morphology of the corolla and androecium and pollen type. Plant habit is not diagnostic but was found to be closely aligned to the phytogeography and ecology of the species within these two clades. Further discussion of results of our morphological survey are presented in the discussion below.

4. Discussion

4.1. Ecology and Biogeography of Monechma Groups I and II

Plants of *Monechma* Group I are slender, annual or perennial herbs with brittle leafy stems. In *J. tetrasperma* and sometimes in *J. eminii*, mature plants can be somewhat shrubby with a woody rootstock but they still retain brittle, slender stems. Species of this clade typically favour open habitats with moderate to high light availability, often in areas that regularly burn and can be considered part of the savanna biome. Several of the species, such as *M. bracteatum*, *M. debile* and *M. monechmoides*, favour disturbed, ruderal habitats although they do not become troublesome weeds. Both the range in growth habit and the favoured habitat types observed in *Monechma* Group I is closely similar to that in *Justicia* sect. *Harnieria* to which *Monechma* Group I is closely allied (see Section 4.2.2 below).

Plants of *Monechma* Group II vary considerably in growth form (see Section 4.2.3), which is again linked closely to ecology. Most species in this clade are perennial herbs or shrublets but *M. ciliatum*, *M. desertorum* and some forms of *M. divaricatum* are annual herbs, though the latter two can be much-branched. Species of the fire-prone savanna biome in the Sudanian and Zambesian phytogeographic regions [29] are typically perennial herbs that produce fertile shoots from a woody base and rootstock that are burnt back during the dry season (similar to geoxylic suffrutices). The exception is *M. ciliatum*, which does not perennate. Species from drier, non-fire prone habitats, particularly in the deserts and xeric shrublands comprising a major extension of the succulent biome in southwest Africa, are most often shrublets with intricate branching (Figure 3).

The majority of species in *Monechma* s.l. occur in the savanna biome and this is reconstructed as the ancestral biome state of the lineage (Figure 5). Our analyses suggest the origin of the lineage at ~31 mya, but the savanna biome is thought to have originated 5–10 mya with the spread and increased dominance of fire-prone C4 grass lineages [67,68]. Indeed, in Africa, phylogenetic evidence suggests that the origin of most lineages of 'underground trees' (geoxylic suffrutices) that place their woody biomass underground to protect it from fire dates to within the last 2 myrs, after the origin and spread of the savanna biome [69]. Either previous studies have grossly inaccurately dated the timing of the origin of the fire-prone savanna biome or *Monechma* s.l. originated in some other biome, with most lineages subsequently shifting to the savanna biome once that biome as we now know it originated; c.f. [70] (Figure 5b). Under the latter scenario, the previous biome(s) occupied by *Monechma* s.l. may have no modern-day analogue, while species in the lineage may have possessed traits that predisposed them to successfully colonise the savanna biome; c.f. [71].

While figuring out the exact timing of colonisation of the savanna biome by *Monechma* s.l. may require further paleobotanical and geological evidence, it seems likely based on our analyses that the ancestors of most extant species were found in savanna except for the conspicuous radiation of >15 species in *Monechma* Group II within the succulent biome (Figure 5). The latter clade, predominantly composed of species in Namibia and neighbouring countries, may have originated as early as ~7 mya (stem age), but seems to have begun substantial diversification within the last one million years (crown age for clade comprising 14 of the 15 species phylogenetically sampled in the clade) (Figure 5). The recency of this radiation is reminiscent of radiation of a distantly related genus of Acanthaceae, *Petalidium*, which similarly has undergone very recent radiation in the succulent biome in southwest Africa, with 39 species originating in the last 0.5 myrs [6]. Indeed, it has been argued that arid environments can facilitate rapid diversification of plant lineages [72], including a suggestion that such has been the case for *Monechma* in this region [24]. These results are surprising, however, as the Namib Desert, which forms the core of the distribution of the succulent biome in southwest Africa, is thought to be among the oldest deserts on Earth, dating to at least 55–80 mya [73]. Clearly, further research, phylogenetic and otherwise, is needed to understand the biogeographical history of this understudied, yet biologically unique region.

4.2. Taxonomic Implications

4.2.1. Morphological and Cytological Traits for the Separation of *Monechma* Groups I and II

Our results confirm that *Monechma* s.l. is a non-monophyletic assemblage of species from two widely separated clades within the Justicioid lineage. In line with earlier studies [13], *Monechma* Group I is nested within the Core *Harnieria* clade whilst *Monechma* Group II is allied to the Diclipterinae clade (Figure 4). However, the constituent species of the two clades of *Monechma*, as revealed by our detailed sampling, do not concur with that interpreted from the limited sampling in earlier studies [13,22].

Specifically, the majority of newly sampled species from tropical Africa with primarily terminal inflorescence spikes bearing the bracts ± highly differentiated from the leaves were not resolved among the *Monechma* Group I clade (Figure 4), as was inferred by those earlier studies [13,22]. Instead, they are resolved in a series of clades within *Monechma* Group II, which is otherwise made up of predominantly southern African species with ± undifferentiated bracts. Thus, the enumeration of *Monechma* Group I in Angola [22] includes taxa from two clades. In light of the present results, the morphological grounds for the separation of the *Monechma* Group I and Group II requires reassessment.

Although quite variable across *Monechma* s.l., growth habit is not diagnostic for separation of these two clades, instead being more closely linked to ecology, which varies particularly within Group II (see Section 4.1) (Figures 5 and 6). Similarly, we do not find traits in the corolla and androecium to be informative. In fact, these traits are remarkably uniform across *Monechma* Group I and II and also across allied taxa in the wider Justicioid lineage, for example in the *Harnieria* and *Tyloglossa* clades. All have the combination of a short and relatively broad corolla tube with tube length usually ≤ the lips (rarely tube > lips e.g., in *M. grandiflorum*) and with prominent transverse ridges ("herring-bone" patterning) on the lower lip (Figure 1); and anthers with offset and often oblique thecae with a prominent pale appendage on the lower theca. In both *Monechma* Groups I and II, the appendage is often bifurcate or even trifurcate at the apex, but this is inconsistent and anthers on the same plant can have entire and divided appendages.

Pollen morphology is often informative in the classification of Acanthaceae [13,14,74], hence pollen morphology has previously been reviewed across the Justicioid lineage [13]. That study found that members of the *Harnieria* clade, including *Monechma* Group I, have bicolporate pollen with each aperture flanked by lines of insulae; see [13] (Figure 11C, D, G and H). The same pollen type was found in the only sampled member of *Monechma* Group II, *M. divaricatum*. We examined one additional species from *Monechma* Group I and two additional species from Group II and further confirm these results (Figure 7). Therefore, pollen type does not distinguish between the two Groups. However, it is noteworthy that this pollen type in Group II is different from other sampled members of Diclipterinae in which the pollen is usually tricolporate hexapseudocolpate, although in *Rhinacanthus virens* (Nees) Milne-Redh. the pollen is tricolporate with insulae; see [13] (Figure 11).

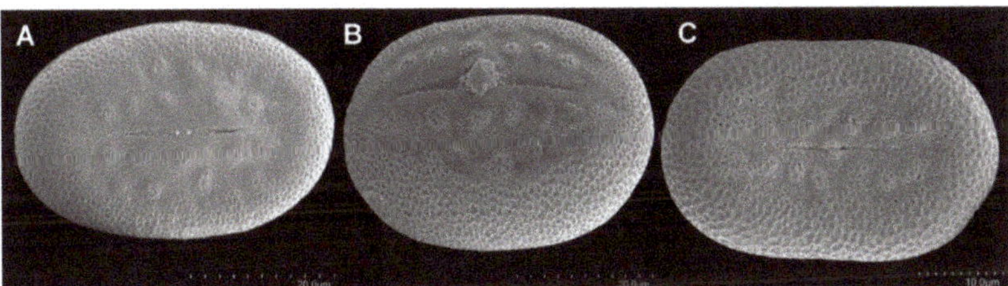

Figure 7. Pollen morphology of *Monechma* Groups I and II. (**A**) and (**B**) *Monechma* Group II: (**A**) *M. leucoderme* (source specimen: *Tripp et al. 2083*, RSA); (**B**) *M. incanum* (*Puff et al. 780416*, RSA). (**C**) *Monechma* Group I: *M. bracteatum* (*Kiel et al. 142*, RSA).

Inflorescence morphology varies considerably within the Justicioid lineage and has been used as an important character in past classification schemes, e.g., [51]. There is considerable variation in inflorescence form in *Monechma* s.l. [19], but based on results presented herein, separation of the two clades using this character is not as straightforward as that proposed in earlier studies [13,22]. In *Monechma* Group I, the flowers are held in axillary or mixed axillary and terminal spikes in which the bracts are highly differentiated from the leaves and are ± broadly elliptic, ovate or obovate (Figure 1A–D). Only in rare cases in Group I is the terminal spike longer than the axillary spikes (or more rarely axillary spikes are absent, e.g., in some specimens of *M. debile*). Each inflorescence unit within the spike can have a single flower, but often contains two or more flowers. The inflorescence arrangement of *J. tetrasperma* has previously been analyzed in detail [19]. That study found that the dichasial units in the proximal portion of the spike have two pairs of bracts, the upper pair of which are highly uneven in size (i.e., three conspicuous bracts and one inconspicuous bract), while units in the distal portion of the spike have only one ± equal pair of bracts. This arrangement is not observed in *Monechma* Group II where the inflorescences are simpler: most species have single-flowered (rarely 2-flowered) inflorescence units per bract, with either one or two inflorescence units per node. In most species from the succulent biome of southern Africa (e.g., *M. cleomoides*, *M. divaricatum* etc.), the bracts are ± undifferentiated from the leaves and so the flowers are axillary (Figure 1K–S,U–X), although, in some species (e.g., *M. genistifolium*, Figure 1T), they form weakly defined terminal spikes. Most species from savanna biome in tropical Africa have bracts that are clearly differentiated from the leaves and have flowers that are held in well-defined terminal spikes, occasionally with additional, usually shorter spikes in the distal-most leaf axils (Figure 1E–I). The only known exception to this is *J. fanshawei*, which has short axillary and terminal spikes (primarily the former). Species of Group II with well-defined spikes typically have bracts that are proportionately narrow and linear or lanceolate. *Justicia kasamae* Vollesen from Zambia (not sampled in our RADseq analysis but included in Group II on the basis of morphology) is an exception, having imbricate bracts that are broad and elliptic to obovate [18].

Capsules of both clades of *Monechma* typically have only two seeds developed due to early abortion of the upper two ovules [19], although all four ovules mature in *J. tetrasperma* of *Monechma* Group I. Seeds are uniform in *Monechma* Group I, being small, 2–3 mm in diameter [17,18], lenticular with a sharp rim and mottled surface, ± symmetrical in cross section and lacking a ridge on one surface, and are glabrous (Figure 8A–C). Seeds of *Monechma* Group II are much more varied in terms of size, shape, surface characteristics and indumentum (Figure 8D–J); this is discussed in more detail in Section 4.2.3. In summary, the seeds of many species in Group II are larger than in Group I and/or they are less strongly compressed with a more rounded rim; they are often asymmetric in a cross section and often have a ± conspicuous longitudinal ridge on one side. Seed colour varies from black to mottled grey or sometimes (e.g., in *M. divaricatum*; Figure 8I) intricately patterned and coloured. Seeds can be glabrous or can have trichomes (Figure 8D,E). Critically, those species of Group II with small, lenticular, glabrous seeds (e.g., *M. desertorum*, Figure 8H) are ± markedly asymmetric in a cross section, with one surface convex and the other often flat or even slightly concave, and have a conspicuous ridge on one side, quite unlike those of Group I.

Chromosome number has also been found to vary considerably across Acanthaceae [75–78]. However, as far as is known, very few chromosome counts are available for *Monechma* s.l. Within *Monechma* Group I, two independent counts of $2n = 28$ have been reported for *Justicia debilis* Lam. (=*M. debile*) [79]. Counts of $2n = 26$ and $2n = 28$ are common in *J.* sect. *Harnieria* to which *Monechma* Group I is closely allied (and also other Old World *Justicia*), but the count within that clade is variable and some species have $2n = 40$ (–50) [76]. Within *Monechma* Group II, a count of $n = 11$ was recorded for *M. ciliatum* [75]. This differs notably from the count of $n = 15$, which is otherwise characteristic of the Diclipterinae clade [13]. Further studies are required to confirm the consistency of the chromosome counts within, and differences between, the two *Monechma* clades.

In summary, the morphological and cytological differences between the two widely separated clades of *Monechma* are subtle and diagnosis is somewhat hindered by significant morphological variation within each clade, particularly among Group II. However, differences in inflorescence form and seed characteristics, potentially together with differences in chromosome number, are informative in separating these two clades. The constituent species for these two clades are listed in Table 3, with those species not sampled in the RADseq phylogeny being placed based on their morphology.

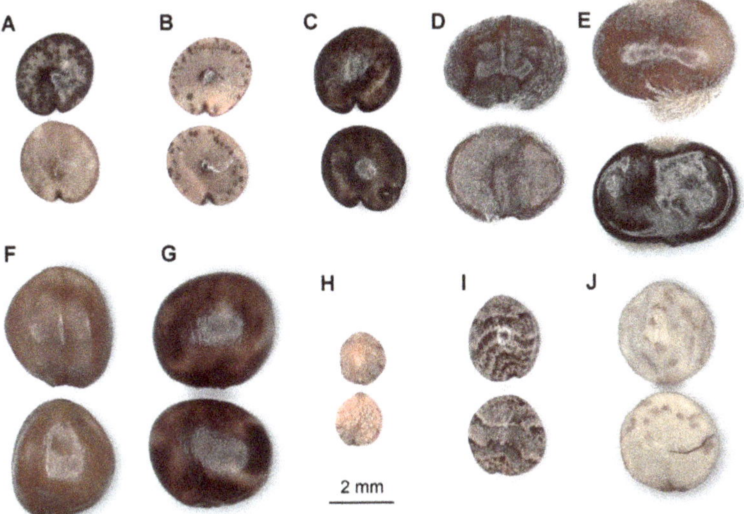

Figure 8. Seeds of *Monechma* Groups I and II. (**A–C**) *Monechma* Group I: (**A**) *M. debile* (source specimen: *Gatheri et al. 79/5*, Kenya); (**B**) *M. bracteatum* (*Kisena 1401*, Tanzania); (**C**) *Justicia eminii* (*Bidgood et al. 3486*, Tanzania). (**D–J**) *Monechma* Group II: (**D**) *M. scabridum* (*Fanshawe F46*, Zambia); (**E**) *M. ciliatum* (*Dere F.H.7047*, Ghana); (**F**) *M. virgultorum* (*Frisby & Maiato 4183*, Angola); (**G**) *J. tricostata* (*Bidgood et al. 3450*, Tanzania); (**H**) *M. desertorum* (*Kolberg & Tholkes HK2038*, Namibia); (**I**) *M. divaricatum* (*Tripp & Dexter 2023*, Namibia); (**J**) *M. mollissimum* (*Salter 1436*, South Africa). All specimens at K.

Table 3. Currently accepted species in *Monechma* s.l., their placement in *Monechma* Groups I and II and their distribution. ˆ denotes that the species has been sampled in the current RADseq phylogeny. Species that were not sampled have been placed based on morphology; placement of these taxa in the clades of *Monechma* Group II should be considered provisional. Combinations in *Monechma* are used wherever available, in preference to combinations in *Justicia*. Geographic range follows TDWG Level 3 codes [65].

Clade	Constituent Species	Distribution
Monechma Group I	ˆ*Monechma bracteatum* Hochst.	Africa: ANG, BOT, ERI, ETH, KEN, MLW, MOZ, NAM, NAT, SOM, SUD, TAN, TVL, UGA, ZAI, ZAM, ZIM; Asia: IND, OMA, YEM
	ˆ*Monechma debile* (Forssk.) Nees	Africa: DJI, ERI, ETH, KEN, SOM, SUD, TAN; Asia: SAU, YEM
	ˆ*Monechma monechmoides* (S. Moore) Hutch.	Africa: ANG, BOT, MLW, MOZ, NAM, TVL, ZAM, ZIM
	Justicia carnosa Hedrén	Africa: SOM
	ˆ*Justicia eminii* Lindau	Africa: BUR, MLW, RWA, TAN, UGA, ZAI, ZAM
	ˆ*Justicia tetrasperma* Hedrén	Africa: TAN, ZAI, ZAM
	ˆ*Justicia* sp. B of Flora Zambesiaca	Africa: MOZ
	Justicia sp. C of Flora Zambesiaca	Africa: ZAM
Monechma Group II		
"*ciliatum/scabridum*" clade	ˆ*Monechma ciliata* (Jacq.) Milne-Redh.	Africa: BEN, BKN, BUR, CAF, CHA, CMN, GAM, GHA, GNB, GUI, ETH, IVO, MLI, MLW, NGA, NGR, RWA, SEN, SIE, SUD, SOSUD, TAN, TOG, UGA, ZAI, ZAM
	ˆ*Monechma depauperatum* (T. Anderson) C.B. Clarke	Africa: BEN, CAF, CMN, GHA, GNB, GUI, IVO, MLI, NGA, SEN, SIE, SOSUD, TOG, ZAI
	ˆ*Monechma ndellense* (Lindau) J. Mlège & Heine	Africa: BKN, CAF, GHA, GUI, MLI, SEN, SUD, TOG

Table 3. *Cont.*

Clade	Constituent Species	Distribution
	^*Monechma scabridum* (S. Moore) C.B. Clarke	Africa: ANG, ZAI, ZAM
	^*Monechma subsessile* (Oliv.) C.B. Clarke	Africa: ANG, BUR, KEN, RWA, TAN, UGA, ZAI, ZAM, ZIM
	^*Justicia attenuifolia* Vollesen	Africa: MOZ, TAN
"*virgultorum*" clade	^*Monechma virgultorum* S. Moore	Africa: ANG
	^*Justicia fanshawei* Vollesen	Africa: ZAM
"*tricostatum*" clade	*Monechma glaucifolium* S. Moore	Africa: ANG
	Monechma lolioides (S. Moore) C.B. Clarke	Africa: ANG
	^*Monechma rigidum* S. Moore	Africa: ANG
	^*Justicia cubangensis* I. Darbysh. & Goyder	Africa: ANG
	Justicia eriniae I. Darbysh.	Africa: ANG
	Justicia laeta S. Moore	Africa: ANG
	^*Justicia tricostata* Vollesen	Africa: TAN, ZAM
"*serotinum*" clade	^*Monechma serotinum* P.G. Mey.	Africa: NAM
"*cleomoides*" clade	^*Monechma australe* P.G. Mey.	Africa: CPP, NAM
	^*Monechma calcaratum* Schinz	Africa: NAM
	Monechma callothamnum Munday	Africa: NAM
	^*Monechma cleomoides* (S. Moore) C.B. Clarke	Africa: ANG, NAM
	Monechma crassiusculum P.G. Mey.	Africa: NAM
	^*Monechma desertorum* (Engl.) C.B. Clarke	Africa: NAM
	^*Monechma divaricatum* (Nees) C.B. Clarke	Africa: ANG, BOT, CPP, CPV, MOZ, NAM, NAT, OFS, SWZ, TVL, ZAM, ZIM
	^*Monechma distichotrichum* (Lindau) P.G. Mey.	Africa: CPP, NAM
	^*Monechma genistifolium* (Engl.) C.B. Clarke	Africa: NAM
	^*Monechma grandiflorum* Schinz	Africa: NAM
	^*Monechma incanum* (Nees) C.B. Clarke	Africa: BOT, CPP, NAM, OFS
	^*Monechma leucoderme* (Schinz) C.B. Clarke	Africa: NAM
	^*Monechma mollissimum* (Nees) P.G. Mey.	Africa: CPP, NAM
	Monechma robustum Bond	Africa: CPP
	^*Monechma salsola* (S. Moore) C.B. Clarke	Africa: ANG, NAM
	Monechma saxatile Munday	Africa: CPP
	^*Monechma spartioides* (T. Anderson) C.B. Clarke	Africa: CPP, NAM
	^*Monechma tonsum* P.G. Mey.	Africa: NAM
Monechma Group II incertae sedis	*Justicia kasamae* Vollesen	Africa: ZAM

4.2.2. Relationship of *Monechma* Group I to *Justicia* sect. *Harnieria*

Our results confirm a close relationship between *Monechma* Group I and *Justicia* sect. *Harnieria* (henceforth sect. *Harnieria*). Sect. *Harnieria* is found to be paraphyletic, with *J. odora* being sister to *Monechma* Group I, although monophyly of sect. *Harnieria* cannot be rejected (p = 0.393; Table 2). This result concurs closely with the findings of earlier studies [13], where a larger sample of species of sect. *Harnieria* was included than in the current study, and where *J. capensis* Thunb. and *J. odora* together were found to be sister to *Monechma* Group I. A number of morphological similarities have been noted between sect. *Harnieria* and *Monechma* s.l. [19], including general corolla shape, presence of conspicuous transverse ridges ("herring-bone" patterning) on a large portion of the lower corolla lip, and biaperturate pollen with insulae, as well as a similar inflorescence form between some members of *Monechma* and sect. *Harnieria*. These similarities, together with the fact that *J. tetrasperma* has an intermediate fruit type, have been used in support of reducing *Monechma* s.l. to a section of *Justicia* [19].

The principle difference between *Monechma* Group I and sect. *Harnieria* is in the fruits. *Monechma* Group I usually have 2-seeded capsules (4-seeded in *J. tetrasperma*) and seeds with a smooth testa. Those of sect. *Harnieria* have 4-seeded capsules with tuberculate seeds, although some species are heterocarpic with highly modified single-seeded indehiscent fruits in addition to the typical dehiscent capsules [13,76].

Variation in sculpturing of the seed testa has been observed within other lineages of Acanthaceae. For example, apparently closely allied members of the genus *Isoglossa* Oerst. in East Africa can have either a rugose testa (e.g., *I. floribunda* C.B. Clarke, *I. grandiflora* C.B. Clarke) or smooth testa (e.g., *I. mbalensis* Brummitt, *I. ufipensis* Brummitt) [80]. Furthermore, within the Justicioid lineage, seeds with a smooth testa are not unique to the two clades of *Monechma*: smooth seeds are observed in several other taxa in *Justicia* s.l. apparently unrelated to the two clades of *Monechma*. These include the group of species *J. grisea* C.B. Clarke, *J. rendlei* C.B. Clarke and *J. salvioides* Milne-Redh. from East Africa, and *J. crebrinodis* Benoist and allies from Madagascar. The *J. crebrinodis* group also have 2-seeded capsules, but are otherwise very different morphologically to the clades of *Monechma* and molecular phylogenetic evidence confirms that they are not closely related [81]. This evidence suggests that variation in seed number and sculpturing may hold only limited taxonomic value at the generic rank within the Justicioid lineage and that it might, therefore, be advisable to treat *Monechma* Group I and sect. *Harnieria* as a single taxonomic unit. Nevertheless, further studies, including more thorough molecular sampling of sect. *Harnieria*, are required to fully decipher relationships within that group and in relation to *Monechma* Group I.

4.2.3. Morphological Variation within *Monechma* Group II

As noted in Section 4.2.1, *Monechma* Group II as recircumscribed here includes a range of morphological variation. Based on the results of the RADseq phylogeny, two major clades are noted (see Results), the latter of which contains several minor clades that can be delimited on morphological grounds and may form the basis for a future classification; these are summarised below. The constituent species for each of these clades are listed in Table 3, with those species not sampled in the RADseq phylogeny being placed based on their morphology.

(i) The *"ciliatum/scabridum"* clade, which contains species of fire-prone savanna biome in tropical Africa, largely associated with the Sudanian and Zambesian phytogeographic regions [29]. These species all share terminal spiciform inflorescences (sometimes with additional spikes in the uppermost leaf axils) and bracts that are ± highly modified from the leaves in both size and shape. The widespread West and Central African species *Monechma ciliatum* is unique in this clade in being an annual herb and in having unusual bristly trichomes on the seeds, restricted to tufts at the apex and base of the seeds, the two tufts being oriented in opposite directions (Figure 8E). All other species in this clade, such as *M. depauperatum*, *M scabridum* and *M. subsessile*, are suffruticose herbs (see Section 4.1). Their seeds are at first finely white-puberulous but later glabrescent. They are rounded or oblate in face view and are compressed but with rounded margins and have one face concave when young, the other face convex and with a ± conspicuous central ridge (Figure 8D). Our data reject the exclusion of *M. ciliatum* from *Monechma* Group II ($p < 0.001$; Table 2).

(ii) The *"virgultorum"* and (iii) the *"tricostatum"* clades, which comprise suffruticose herbs of southern tropical Africa, mainly associated with the Zambesian phytogeographic region [29]. As in the *"ciliatum/scabridum"* clade, they have predominantly terminal spiciform inflorescences with highly modified bracts, the exception being *Justicia fanshawei*, which has short axillary and terminal spikes. The seeds in these clades are less compressed than in the *"ciliatum/scabridum"* clade and are glabrous (Figure 8F,G). Several members of these two clades have secund inflorescence spikes in which only one of each pair of bracts is fertile, but this is not universal, for example both *M. rigidum* and *J. tricostata* can have opposite flowers along the spike. The *"tricostatum"* clade differs from the *"virgultorum"* clade in having prominently 3-veined calyx lobes, bracts and bracteoles, the veins often being a markedly different colour from the intercostal surfaces. In *M. virgultorum* and *M. fanshawei*, the calyces are at most only weakly 3-veined with only the midvein ever prominent.

(iv) The *"serotinum"* clade. *Monechma serotinum*, a rare species endemic to the Kaokoveld of Namibia, occupies a position in the phylogeny between the tropical African, fire-prone savanna clades outlined above and the group of species that are concentrated in the non-fire prone deserts and bushlands of southern Africa, i.e., the *"cleomoides"* clade discussed below. *Monechma serotinum* is also somewhat intermediate in morphological terms. It has a well-defined lax terminal spike with reduced bracts in comparison to the leaves as in most members of the savanna clades, but it has the dwarf shrubby habit of many species of the *"cleomoides"* clades (see below), in keeping with its non-fire prone habitat. The seeds of this species are glabrous, compressed, and asymmetric in cross section, with one face convex.

(v) The "*cleomoides*" clade. The remainder of the taxa in *Monechma* Group II are included in a single, large clade which comprises southern African taxa of dry, non-fire prone habitats including deserts and bushlands of the succulent biome. Most of the species are dwarf shrublets, often intricately branched and sometimes with gnarled lignified mature branches, although *M. desertorum* and some forms of *M. divaricatum* are annual herbs [20]. These species are united by having single-flowered axillary inflorescences with the bracts undifferentiated or not markedly differentiated from the leaves, although in some species such as *M. genistifolium* the flowers can together form ill-defined leafy terminal spikes. Many species have a complex indumentum comprising multiple trichome types (often both eglandular and glandular), and in some taxa the trichomes can be branched; for example, *M. incanum* has biramous trichomes on the vegetative parts and *M. calcaratum* has stellate trichomes on the stems [20]. The calyx lobes in this clade are either prominently single-veined or the venation is obscure. *Monechma divaricatum* is notable for having only four calyx lobes with no evidence of a vestigial fifth lobe; all other species in this clade (and elsewhere in *Monechma* Group II) have five-lobed calyces. The seeds in this clade are always glabrous, usually small and compressed, with either a rounded or sharp rim and ± asymmetric in cross section, with one face being more convex than the other and often having a prominent central ridge (Figure 8H–J).

The "*cleomoides*" clade is notable for containing several taxonomically challenging taxa, particularly regarding three highly variable aggregate species: *M. cleomoides*, *M. divaricatum* and *M. spartioides*. We sampled multiple accessions of the former two species. Whilst *M. divaricatum* is monophyletic, albeit with significant phylogenetic diversity, *M. cleomoides* is resolved as polyphyletic. Three accessions of that species are resolved in a clade that also contains two accessions of *M. tonsum*. These two taxa are separated primarily by differences in indumentum: *M. tonsum* has a short velvety indumentum whilst that of *M. cleomoides* usually includes ± dense mixed short and long shining trichomes (Figure 1U–W). Our results suggest that this difference in indumentum may be of limited taxonomic significance. A fourth accession of *M. cleomoides* (*Klaassen et al.* 2530) is resolved as sister to a clade containing *M. genistiifolium* and *M. australe*, which is difficult to reconcile with the morphological evidence, in view of the fact that these two species are morphological dissimilar to *M. cleomoides*. An AU test, however, rejects the monophyly of *M. cleomoides* and *M. tonsum* in addition to a monophyletic *M. cleomoides* + *M. tonsum* relationship ($p < 0.001$; Table 2).

4.2.4. The Status of *Monechma varians* and *Justicia platysepala*

Monechma varians is a rare species, confined to the Nyika Plateau of Malawi. It was recently transferred to *Justicia* sect. *Monechma* [18], although with a note that the capsule and seeds of this species had not been seen. The RADseq data place *M. varians* outside either of the two "*Monechma*" clades, it instead being resolved as sister to *J. kirkiana* in the *Tyloglossa* clade. A specimen of *M. varians* at K, *Synge WC437*, was annotated by M. Hedrén in 2000, stating "a *Justicia* close to *J. linearispica* C.B. Cl[arke]. Capsule probably 4-seeded, inflorescences as in *linearispica*". We concur with this suggestion as these two species are morphologically similar, and earlier molecular phylogenetic studies have placed *Justicia linearispica* within the *Tyloglossa* clade [13].

Justicia platysepala was originally described in *Monechma* on the basis of it having two- (or one-) seeded capsules. However, the seeds of this species—together with the related species *J. guerkeana* Schinz, also previously described in *Monechma* as *M. clarkei* Schinz—are tuberculate, and thus quite unlike those of *Monechma* s.l. Our results confirm that *J. platysepala* does not belong within either of the two clades of *Monechma*.

5. Conclusions

The findings of this study confirm that the genus *Monechma* (or *Justicia* sect. *Monechma*), as previously circumscribed, represents two widely separated clades. Our findings provide insights into the evolutionary histories of these two clades. Particularly striking is the relatively recent radiation (diversifying ca. 1.9 mya) in *Monechma* Group II into the ancient deserts and xeric shrublands of the succulent biome in southwest Africa. While colonisation of the succulent biome may have involved relaxed selection on traits required to survive regular fires that are present in the savanna biome, it required adaptation to higher water deficits (evidenced by lower precipitation throughout the year) and greater extremes of low and high temperatures (higher annual temperature range). Clearly, this clade in *Monechma* Group II was able to adapt, as the radiation now accounts for more than half of the current species diversity in *Monechma* Group II despite the much longer evolutionary history of the clade within the savanna biome of tropical Africa. Our results support the need for future research to further understand the biogeographical history of these centers of biodiversity in Africa.

Given that *Justicia* s.l. is highly paraphyletic and encompasses all the taxa within the Justicioid lineage, and given the desire to avoid losing valuable taxonomic information that would be incurred through an all-encompassing *Justicia*, the only plausible option is to recognise distinct clades within *Justicia* s.l. as discrete genera and to seek morphological characters in support of these segregations. Our results show that *Monechma* Groups I and II are distinguishable from one another, albeit subtly so, by differences in the inflorescence structure and seed morphology. These two clades should therefore be elevated to generic status, and a forthcoming study will address the nomenclatural implications of recognising them as separate genera. Detailed studies employing NGS techniques, comparable to that presented in the current work, are required across the Justicioid lineage in order to delimit other genera in this complex but ecologically important plant group.

Supplementary Materials: The following are available online at http://www.mdpi.com/1424-2818/12/6/237/s1, Figure S1: Summary phylogeny of the majority-rule consensus tree from Bayesian analysis illustrating the 10 major clades of the Justicioid lineage from [13]. Within the *Harnieria* clade, species of *Monechma* Group I are sister to *Justicia odora*. Embedded within the Diclipterinae clade, species of *Monechma* Group II are sister to *Kenyacanthus ndorensis* + (*Hypoestes* + *Dicliptera*). Thickened branches are supported by ≥0.98 Bayesian posterior probability and ≥70% maximum likelihood bootstrap. Size of clades corresponds to the number of taxa sampled in each clade, Figure S2: The most likely phylogenetic hypothesis generated from ddRAD-seq loci from the iPYRAD de novo assembly with the minimum sample to retain a locus set to four. Asterisks [*] indicate 100% ML bootstrap and dashes [-] indicate <70% ML bootstrap, Figure S3: The most likely phylogenetic hypothesis generated from ddRAD-seq loci from the iPYRAD de novo assembly with the minimum sample to retain a locus set to 10. Asterisks [*] indicate 100% ML bootstrap and dashes [-] indicate <70% ML bootstrap, Figure S4: The most likely phylogenetic hypothesis generated from ddRAD-seq loci from the iPYRAD de novo assembly with the minimum sample to retain a locus set to 30. Asterisks [*] indicate 100% ML bootstrap and dashes [-] indicate <70% ML bootstrap, Figure S5: Phylogenetic relationships among the samples included in our study based on quartet multispecies coalescent analyses of loci resulting from the iPYRAD assembly. Numbers at nodes represent percent support across 1000 replicate quartet analyses. Asterisks [*] indicate 100% support, Table S1: Taxon, source of plant material, number of ddRAD reads per sample and number of loci per sample in our final assembly. Taxa are listed in alphabetical order by genus and species. Table S2: Results of tests for the best-fit model of evolution for biome in the ancestral state reconstruction analyses. The model in bold was selected.

Author Contributions: Conceptualization, I.D., C.A.K. and E.A.T.; methodology, C.A.K., E.A.T., I.D. and K.G.D.; software, C.A.K., E.A.T. and K.G.D.; validation, C.A.K., I.D., E.A.T. and K.G.D.; formal analysis, C.A.K., C.M.A. and E.A.T.; investigation, I.D., C.A.K., E.A.T. and K.G.D.; resources, C.A.K., E.A.T. and I.D.; data curation, C.A.K., C.M.A., E.A.T. and I.D.; writing—original draft preparation, I.D., C.A.K. and C.M.A.; writing—review and editing, I.D., C.A.K., E.A.T., F.M.C. and K.G.D.; visualization, C.A.K., I.D. and K.G.D.; supervision, C.A.K. and E.A.T.; project administration, I.D., C.A.K. and E.A.T.; funding acquisition, C.A.K. and E.A.T. All authors have read and agreed to the published version of the manuscript.

Funding: This research was funded by the U.S. National Science Foundation, grant number DEB 1754845 to C.A.K. and E.A.T., and Rancho Santa Ana Botanic Garden. Support for field collection in Namibia and Angola was received from the U.S. National Science Foundation grant numbers DEB 0919594 and DEB 1354964 to E.A.T. Support for research visits to WIND herbarium and field collection in Namibia by I.D. and E.A.T., was received from the Southern African Science Service Centre for Climate Change and Adaptive Land Management (SASSCAL) programme, Task 060: "Establish and Improve baseline inventories for spatial data and biodiversity—Flora of Namibia Project".

Acknowledgments: Fieldwork in Namibia by E.A.T. and K.D. and by I.D. was facilitated by Ezekeil Kwembeya and Esmerialda Klaassen (WIND herbarium, National Botanical Research Institute) and in Angola by Fernanda Lages and Francisco Maiato (LUBA Herbarium, ISCED-Huíla). C.A.K. thanks Mathais Mbale, Itambo Malombe, Jeffery Morawetz, Jonathan Ogweno and Quentin Luke for assisting with field studies. We are grateful to the curators of the following herbaria for access to and/or permission to sample specimens: CAS, COLO, EA, J, K, and RSA. We thank Stefan Dressler and Marco Schmidt of the Forschungsinstitut Senckenberg for facilitating access to field photographs of *Monechma* species. We also want to thank Biofrontiers lab at UC Boulder for their help with troubleshooting library preparations.

Conflicts of Interest: The authors declare no conflict of interest. The funders had no role in the design of the study; in the collection, analyses, or interpretation of data; in the writing of the manuscript, or in the decision to publish the results.

References

1. Darbyshire, I.; Kordofani, M.; Farag, I.; Candiga, R.; Pickering, H. *The Plants of Sudan and South Sudan: An Annotated Checklist*; Royal Botanic Gardens, Kew: Richmond, UK, 2015.

2. Darbyshire, I.; Timberlake, J.; Osborne, J.; Rokni, S.; Matimele, H.; Langa, C.; Datizua, C.; de Sousa, C.; Alves, T.; Massingue, A.; et al. The endemic plants of Mozambique: Diversity and conservation status. *PhytoKeys* **2019**, *136*, 45–96. [CrossRef] [PubMed]

3. Onana, J.M. The Vascular Plants of Cameroon. A taxonomic checklist with IUCN assessments. In *Flore du Cameroon*; National Herbarium of Cameroon: Yaoundé, Cameroon, 2011; Volume 39.

4. Gosline, G.; Cheek, M.; Bidault, E.; van der Burgt, X.M.; Challen, G.; Couch, C.; Couvreur, T.; Darbyshire, I.; Dawson, S.; Goyder, D.; et al. Known plant diversity of the Republic of Guinea increased by 30% in a new taxonomically curated checklist. Unpublished work.

5. Vollesen, K. *Blepharis (Acanthaceae): A Taxonomic Revision*; Royal Botanic Gardens, Kew: Richmond, UK, 2000.

6. Tripp, E.A.; Tsai, Y.H.E.; Zhuang, Y.; Dexter, K. RADseq dataset with 90% missing data fully resolves recent radiation of *Petalidium* (Acanthaceae) in the ultra-arid deserts of Namibia. *Ecol. Evol.* **2017**, *7*, 7920–7936. [CrossRef] [PubMed]

7. Darbyshire, I.; Tripp, E.A.; Chase, F.M. A taxonomic revision of Acanthaceae tribe Barlerieae in Angola and Namibia. Part 1. *Kew Bull.* **2019**, *74*, 1–85. [CrossRef]

8. Darbyshire, I. Taxonomic notes and novelties in the genus *Isoglossa* (Acanthaceae) from east Africa. *Kew Bull.* **2009**, *64*, 401–427. [CrossRef]

9. Tripp, E.A.; Dexter, K.G. Taxonomic novelties in Namibian *Ruellia* (Acanthaceae). *Syst. Bot.* **2012**, *37*, 1023–1030. [CrossRef]

10. Darbyshire, I.; Luke, Q. *Barleria mirabilis* (Acanthaceae): A remarkable new tree species from west Tanzania. *Kew Bull.* **2016**, *71*, 13. [CrossRef]

11. McDade, L.A.; Daniel, T.F.; Kiel, C.A. Towards a comprehensive understanding of phylogenetic relationships among lineages of Acanthaceae *s.l.* (Lamiales). *Am. J. Bot.* **2008**, *95*, 1136–1152. [CrossRef] [PubMed]

12. McDade, L.A.; Daniel, T.F.; Kiel, C.A. The *Tetramerium* Lineage (Acanthaceae, Justicieae) revisited: Phylogenetic relationships reveal polyphyly of many New World genera accompanied by rampant evolution of floral morphology. *Syst. Bot.* **2018**, *43*, 97–116. [CrossRef]

13. Kiel, C.A.; Daniel, T.F.; Darbyshire, I.; McDade, L.A. Unraveling relationships in the morphologically diverse and taxonomically challenging "justicioid" lineage (Acanthaceae: Justicieae). *Taxon* **2017**, *66*, 645–674. [CrossRef]

14. Kiel, C.A.; Daniel, T.F.; McDade, L.A. Phylogenetics of New World 'justicioids' (Justicieae: Acanthaceae): Major lineages, morphological patterns, and widespread incongruence with classification. *Syst. Bot.* **2018**, *43*, 459–484. [CrossRef]

15. Darbyshire, I.; Kiel, C.A.; Daniel, T.F.; McDade, L.A.; Luke, W.R.Q. Two new genera of Acanthaceae from tropical Africa. *Kew Bull.* **2019**, *74*, 1–25. [CrossRef]

16. Tripp, E.A.; Darbyshire, I.; Daniel, T.F.; Kiel, C.A.; McDade, L.A. Revised classification of Acanthaceae including treatment of previously unplaced genera and worldwide dichotomous keys. Unpublished work.

17. Vollesen, K. *Justicia* . In *Flora of Tropical East Africa*; Beentje, H., Ed.; Royal Botanic Gardens, Kew: Richmond, UK, 2010; pp. 495–601.

18. Vollesen, K. *Justicia* . In *Flora Zambesiaca*; Timberlake, J.R., Martins, E.S., Eds.; Royal Botanic Gardens, Kew: Richmond, UK, 2015; Volume 8, pp. 162–224.

19. Hedrén, M. *Justicia tetrasperma* sp. nov.: A linking species between *Justicia* and *Monechma* (Acanthaceae). *Nordic J. Bot.* **1990**, *10*, 149–153. [CrossRef]

20. Munday, J. *Monechma* . In *Flora of Southern Africa*; Leistner, O.A., Ed.; National Botanical Institute: Pretoria, South Africa, 1995; Volume 30, pp. 47–61.

21. Hedberg, I.I.; Ensermu Kelbessa, E.S.; Sebsebe Demissew, P.E. (Eds.) Ensermu Kelbessa Acanthaceae. In *Flora of Ethiopia & Eritrea*; The National Herbarium, Addis Ababa University: Addis Ababa, Ethiopia; The Department of Systematic Botany, Uppsala University: Uppsala, Sweden, 2006; Volume 5, pp. 345–495.

22. Darbyshire, I.; Goyder, D.J. Notes on *Justicia* sect. *Monechma* (Acanthaceae) in Angola, including two new species. *Blumea* **2019**, *64*, 97–107. [CrossRef]

23. Dressler, S.; Schmidt, M.; Zizka, G. *African Plants—A Photo Guide*; Forschungsinstitut Senckenberg: Frankfurt/Main, Germany, 2014. Available online: www.africanplants.senckenberg.de (accessed on 26 February 2020).

24. Munday, J. The distribution of *Monechma* (Acanthaceae) species in southern Africa. *Bothalia* **1983**, *14*, 575–578. [CrossRef]

25. Ringelberg, J.J.; Zimmermann, N.E.; Weeks, A.; Lavin, M.; Hughes, C.E. Biomes as evolutionary arenas: Convergence and conservatism in the trans-continental succulent biome. *Glob. Ecol. Biogeogr.* **2020**. [CrossRef]

26. Schrire, B.D.; Lavin, M.A.T.T.; Lewis, G.P. Global distribution patterns of the Leguminosae: Insights from recent phylogenies. *Biol. Skr.* **2005**, *55*, 375–422.

27. Klaassen, E.; Kwembeya, E. (Eds.) *A Checklist of Namibian Indigenous and Naturalised Plants*; Occasional Contributions No. 5; National Botanical Research Institute: Windhoek, Namibia, 2013.

28. Chase, F.M. Flora of Namibia: *Monechma*. Unpublished work.

29. White, F. *Vegetation of Africa: A Descriptive Memoir to Accompany the UNESCO/AETFAT/UNSO Vegetation Map of Africa*; UNESCO: Paris, France, 1983.

30. Daniel, T.F.; Tripp, E.A. *Louteridium* (Acanthaceae: Acanthoideae: Ruellieae: Trichantherinae): Taxonomy, phylogeny, reproductive biology, and conservation. *Proc. Calif. Acad. Sci. Ser. 4* **2018**, *65*, 41–106.

31. Tripp, E.A.; Darbyshire, I. *Mcdadea*: A new genus of Acanthaceae endemic to the Namib Desert of Southwestern Angola. *Syst. Bot.* **2020**, *45*, 200–211. [CrossRef]

32. Comito, R.P. A RADseq Phylogeny of *Barleria* (Acanthaceae) Resolves Fine-Scale Relationships. Master's Thesis, California State University, Long Beach, CA, USA, 2019.

33. Kiel, C.A.; Tripp, E.A.; Fisher, A.E.; McDade, L.A. Is pollen involved too? Floral traits and pollinators in New World *Justicia* (Acanthaceae). In Proceedings of the Conference Presentation, Botany 2019 Conference, Tucson, AZ, USA, 27–31 July 2019. Available online: https://2019.botanyconference.org/engine/search/index.php?func=detail&aid=770 (accessed on 30 April 2020).

34. Doyle, J.J.; Doyle, J.L. A rapid DNA isolation procedure for small amounts of fresh leaf tissue. *Phytochem. Bull.* **1987**, *19*, 11–15.

35. Parchman, T.L.; Gompert, Z.; Mudge, J.; Schilkey, F.D.; Benkman, C.W.; Buerkle, C.A. Genome-wide association genetics of an adaptive trait in lodgepole pine. *Mol. Ecol.* **2012**, *21*, 2991–3005. [CrossRef] [PubMed]

36. Andrews, S. *FastQC: A Quality Control Tool for High Throughput Sequence Data*; Babraham Institute: Cambridge, UK, 2017. Available online: https://www.bioinformatics.babraham.ac.uk/projects/fastqc/ (accessed on 26 February 2020).

37. Eaton, D.A.R. PyRAD: Assembly of *de novo* RADseq loci for phylogenetic analysis. *Bioinformatics* **2014**, *30*, 1844–1849. [CrossRef] [PubMed]

38. Eaton, D.A.R.; Overcast, I. ipyrad: Interactive assembly and analysis of RADseq datasets. *Bioinformatics* **2020**. [CrossRef] [PubMed]

39. Nguyen, L.-T.; Schmidt, H.A.; von Haeseler, A.; Minh, B.Q. IQ-TREE: A fast and effective stochastic algorithm for estimating maximum likelihood phylogenies. *Mol. Biol. Evol.* **2015**, *32*, 268–274. [CrossRef] [PubMed]

40. Darriba, D.; Posada, D.; Kozlov, A.M.; Stamatakis, A.; Benoit, M.; Flouri, T. ModelTest NG: A new and scalable tool for the selection of DNA and protein evolutionary models. *Mol. Biol. Evol.* **2020**, *37*, 291–294. [CrossRef] [PubMed]

41. Eaton, D.A.R.; Spriggs, E.L.; Park, B.; Donoghue, M.J. Misconceptions on missing data in RAD-seq phylogenetics with a deep-scale example from flowering plants. *Syst. Biol.* **2017**, *66*, 399–412. [CrossRef] [PubMed]

42. Chifman, J.; Kubatko, L. Quartet inference from SNP data under the coalescent model. *Bioinformatics* **2014**, *30*, 3317–3324. [CrossRef] [PubMed]

43. Shimodaira, H. An approximately unbiased test of phylogenetic tree selection. *Syst. Biol.* **2002**, *51*, 492–508. [CrossRef] [PubMed]

44. Maddison, W.P.; Maddison, D.R. *Mesquite: A Modular System for Evolutionary Analysis*, Version 2.72; 2009. Available online: http://mesquiteproject.org (accessed on 26 February 2020).

45. Paradis, E. Molecular dating of phylogenies by likelihood methods: A comparison of models and a new information criterion. *Mol. Phylogenet. Evol.* **2013**, *67*, 436–444. [CrossRef] [PubMed]

46. Paradis, E.; Claude, J.; Strimmer, K. APE: Analyses of Phylogenetics and Evolution in R language. *Bioinformatics* **2004**, *20*, 289–290. [CrossRef] [PubMed]

47. R Core Team. *R: A Language and Environment for Statistical Computing*; R Foundation for Statistical Computing: Vienna, Austria, 2019. Available online: http://www.R-project.org/ (accessed on 15 April 2020).

48. Smith, S.A.; O'Meara, B.C. treePL: Divergence time estimation using penalized likelihood for large phylogenies. *Bioinformatics* **2012**, *28*, 2689–2690. [CrossRef] [PubMed]

49. Tripp, E.A.; McDade, L.A. A rich fossil record yields calibrated phylogeny for Acanthaceae (Lamiales) and evidence for marked biases in timing and directionality of intercontinental disjunctions. *Syst. Biol.* **2014**, *63*, 660–684. [CrossRef] [PubMed]

50. Mautino, L.R. Nuevas especies de palinomorfos de las formaciones San José y Chiquimil (Mioceno Medio y Superior), noroeste de Argentina. *Rev. Bras. Paleontol.* **2011**, *14*, 279–290. [CrossRef]

51. Graham, V.A.W. Delimitation and infra-generic classification of *Justicia* (Acanthaceae). *Kew Bull.* **1988**, *43*, 551–624. [CrossRef]

52. Burgess, N.; Hales, J.A.; Underwood, E.; Dinerstein, E.; Olson, D.; Itoua, I.; Schipper, J.; Ricketts, T.; Newman, K. *Terrestrial Ecoregions of Africa and Madagascar: A Conservation Assessment*; Island Press: Washington, DC, USA, 2004.

53. Beaulieu, J.M.; O'Meara, B.C.; Donoghue, M.J. Identifying hidden rate changes in the evolution of a binary morphological character: The evolution of plant habit in campanulid angiosperms. *Syst. Biol.* **2013**, *62*, 725–737. [CrossRef] [PubMed]

54. Pagel, M. Inferring evolutionary processes from phylogenies. *Zool. Scr.* **1997**, *26*, 331–348. [CrossRef]

55. Lewis, P.O. A likelihood approach to estimating phylogeny from discrete morphological character data. *Syst. Biol.* **2001**, *50*, 913–915. [CrossRef] [PubMed]

56. Posada, D.; Buckley, T.R. Model selection and model averaging in phylogenetics: Advantages of Akaike Information Criterion and Bayesian approaches over likelihood ratio tests. *Syst. Biol.* **2004**, *53*, 793–808. [CrossRef] [PubMed]

57. WorldClim. Maps, Graphs, Tables, and Data of the Global Climate. Available online: http://worldclim.org (accessed on 15 April 2020).

58. Fick, S.E.; Hijmans, R.J. WorldClim 2: New 1-km spatial resolution climate surfaces for global land areas. *Int. J. Climatol.* **2017**, *37*, 4302–4315. [CrossRef]

59. Hijmans, R.J. *Raster: Geographic Data Analysis and Modeling R Package*; 2014. Available online: https://rdrr.io/cran/raster/ (accessed on 15 April 2020).

60. Revell, L.J. phytools: An R package for phylogenetic comparative biology (and other things). *Methods Ecol. Evol.* **2012**, *3*, 217–223. [CrossRef]

61. Felsenstein, J. Phylogenies and the comparative method. *Am. Nat.* **1985**, *125*, 1–15. [CrossRef]

62. Revell, L.J. Two new graphical methods for mapping trait evolution on phylogenies. *Methods Ecol. Evol.* **2013**, *4*, 754–759. [CrossRef]

63. Thiers, B. Index Herbariorum: A Global Directory of Public Herbaria and Associated Staff. New York Botanical Garden's Virtual Herbarium. Continuously Updated. Available online: http://sweetgum.nybg.org/science/ih/ (accessed on 26 February 2020).

64. JSTOR Global Plants. Available online: https://plants.jstor.org/ (accessed on 26 February 2020).

65. Brummitt, R.K. *World Geographical Scheme for Recording Plant Distributions*, 2nd ed.; International Working Group on Taxonomic Databases for Plant Sciences (TDWG), Hunt Institute for Botanical Documentation, Carnegie Mellon University: Pittsburgh, PA, USA, 2001. Available online: http://www.tdwg.org/standards/109 (accessed on 26 February 2020).

66. Immelman, K.L. *Justicia* . In *Flora of Southern Africa*; Leistner, O.A., Ed.; National Botanical Institute: Pretoria, South Africa, 1995; Volume 30, Part 3; pp. 18–46.

67. Beerling, D.J.; Osborne, C.P. The origin of the savanna biome. *Glob. Chang. Biol.* **2006**, *12*, 2023–2031. [CrossRef]

68. Edwards, E.J.; Osborne, C.P.; Strömberg, C.A.; Smith, S.A.; C4 Grasses Consortium. The origins of C4 grasslands: Integrating evolutionary and ecosystem science. *Science* **2010**, *328*, 587–591. [CrossRef] [PubMed]

69. Maurin, O.; Davies, T.J.; Burrows, J.E.; Daru, B.H.; Yessoufou, K.; Muasya, A.M.; Van der Bank, M.; Bond, W.J. Savanna fire and the origins of the 'underground forests' of Africa. *New Phytol.* **2014**, *204*, 201–214. [CrossRef] [PubMed]

70. Eiserhardt, W.L.; Couvreur, T.L.; Baker, W.J. Plant phylogeny as a window on the evolution of hyperdiversity in the tropical rainforest biome. *New Phytol.* **2017**, *214*, 1408–1422. [CrossRef] [PubMed]

71. Ackerly, D.D. Adaptation, niche conservatism, and convergence: Comparative studies of leaf evolution in the California chaparral. *Am. Nat.* **2004**, *163*, 654–671. [CrossRef] [PubMed]

72. Stebbins, G.L. Aridity as a stimulus to plant evolution. *Am. Nat.* **1952**, *86*, 33–44. [CrossRef]

73. Goudie, A. Namib Sand Sea: Large dunes in an ancient desert. In *Geomorphological Landscapes of the World*; Migon, P., Ed.; Springer: Dordrecht, Germany, 2009; pp. 163–169.

74. Scotland, R.W.; Vollesen, K. Classification of Acanthaceae. *Kew Bull.* **2000**, *55*, 513–589. [CrossRef]

75. Miège, J.; Josserand, N. Nombres chromosomiques d'èspeces africaines et malgaches. *Candollea* **1972**, *27*, 283–292.

76. Hedrén, M. *Justicia* sect. *Harnieria* (Acanthaceae) in tropical Africa. *Acta Univ. Upsal. Symb. Bot. Upsal.* **1989**, *29*, 1–141.

77. Daniel, T.F.; Chuang, T.I. Chromosome numbers of New World Acanthaceae. *Syst. Bot.* **1993**, *18*, 283–289. [CrossRef]

78. Daniel, T.F. Chromosome numbers of miscellaneous Malagasy Acanthaceae. *Brittonia* **2006**, *58*, 291–300. [CrossRef]

79. Grant, W.F. A cytogenetic study in the Acanthaceae. *Brittonia* **1955**, *8*, 121–149. [CrossRef]

80. Darbyshire, I.; Vollesen, K. Ensermu Kelbessa *Acanthaceae (Part 2)*. In *Flora of Tropical East Africa*; Beentje, H.J., Ed.; Royal Botanic Gardens, Kew: Richmond, UK, 2010.

81. McDade, L.A.; Daniel, T.F.; Darbyshire, I.; Kiel, C.A. Justicieae II: Resolved placement of many genera and recognition of a new lineage sister to Isoglossinae. Unpublished work.

Article

Diversity of Tree Species in Gap Regeneration under Tropical Moist Semi-Deciduous Forest: An Example from Bia Tano Forest Reserve

Maame Esi Hammond * and Radek Pokorný

Department of Silviculture, Faculty of Forestry and Wood Technology, Mendel University in Brno, Zemědělská 3, 61300 Brno, Czech Republic; radek.pokorny@mendelu.cz
* Correspondence: xhammon1@mendelu.cz; Tel.: +420-773-928-506

Received: 16 June 2020; Accepted: 31 July 2020; Published: 1 August 2020

Abstract: In a quest to improve the diversity and conservation of native tree species in tropical African forests, gap regeneration remains all-important nature-promoting silviculture practice and ecosystem-based strategy for attaining these ecological goals. Nine gaps of varying sizes (286–2005 m^2) were randomly selected: three each from undisturbed, slightly disturbed and disturbed areas within Bia Tano Forest Reserve of Ghana. Within individual gaps, four transects (North–South–East–West directions) followed by 10 subsampling regions of 1 m^2 at 2 m apart were established along each transect. Data showed 63 tree species from 21 families in the study. Although, all estimated diversity indices showed significant biodiversity improvements in all gaps at $p < 0.05$ level. Yet, there were no significant variations amongst gaps. Additionally, tree species differed between gaps at the undisturbed and the two disturbance-graded areas while no differences were presented between disturbance-graded areas. Balanced conservation between Green Star and Reddish Star species and imbalanced conservation between Least Concern, Near Threatened and Vulnerable species in the International Union for Conservation of Nature (IUCN) Red List were found, showing the reserve's long-term prospects for economic and ecological benefits of forest management. Thus, there is a need for higher priority for intensive management to regulate various anthropogenic disturbances so as to protect the biological legacies of the reserve.

Keywords: Bia Tano Forest Reserve; conservation; gap; regeneration; species composition; species diversity

---

1. Introduction

Tropical forests are so endowed with beautiful vegetation that often display a great diversity of tree species including other life forms. Tropical forests mostly present a broad spectrum of optimal growing habitats that have ecologically tolerable niches for the regeneration of diversified tree species [1,2]. Globally, tropical forests are recognized as the most extensive terrestrial biodiversity [2–4], with the tropical moist forests forming an integral component of the global natural forests [5]. So, understanding stand composition and structure is a useful silviculture knowledge in evaluating species conservation, forest management and forest sustainability [6,7]. Unfortunately, most tropical forests are subjects of various environmental and anthropogenic pressures [2,8–10], and thus, if not managed properly, they may pose adverse effects on floral biodiversity in the future. Therefore, silviculture and forest management interventions which are beneficial to the maintenance of the overall biodiversity, productivity and sustainability of tropical forests are needful [11]. For sustainable tropical forest management, natural regeneration is crucial for the preservation and maintenance of biodiversity because it contributes substantially to tree species dynamics and composition within several tropical forest ecosystems [6,12,13]. Nevertheless, natural regeneration is best enhanced within the gap environment [14–16].

Gap regeneration could be defined as the natural regeneration prevailing under forest canopy gaps. Gaps are openings within forest canopies formed through death(s) or injury(s) of one or more trees in a stand [14,17]. Gap formation is the single most critical event which defines the composition and spatial structure of tropical forests [8] because of its frequent occurrences in the tropics [10,14,18]. The process of gap formation promotes mechanisms for species richness by offering regeneration microsites that have niche diversification along the gap-created light gradient [14,16] for the empowerment of tree species coexistence, and thus, species diversity [18,19]. Furthermore, gaps remain a decisive phase in natural regeneration because they generally determine what would regenerate and also drive the floristic composition of the whole forest cycle following forest disturbances [14,16]. Primarily, species composition within gaps is significantly determined by the chance of gap formation rather than regeneration niches [19,20]. Moreover, the floristic nature of forest communities can predict the regeneration success of tree species depending on the ecological characteristics at growing sites, species diversity and regeneration status of species [6,12]. Notwithstanding, the diagnostic combinations of vegetation characteristics (forest types and vegetation architecture), gap characteristics, plant functional traits and site conditions [21] factors comprehensively influence tree regeneration in gaps. However, the responses of natural regeneration of different tree species in gaps are different because of individual species ecological requirements for growth, particularly concerning their differential light regimes [19,21]. Subsequently, this leads to the organization of tree species with different physiological light-specific demands for establishment and growth in gaps [15,16].

Due to the long-standing exploitation and trade of economic timber products with their bulk quantities being harvested from natural tropical moist forests [9]. Logging operations continue to be more intensified pronouncedly in the semi-deciduous zones than in any other forest type in Ghana due to the hosts of higher densities of economically desirable timber tree species [9,22]. Considering the economic benefits tropical moist semi-deciduous forest offers at both local and global fonts coupled with its immeasurable significant ecological contribution specifically towards global plant biodiversity conservation. It is, therefore, a necessity to strategically channel attention to this positive effect of logging: gap formation on the maintenance of biodiversity extant, which particularly champions restoration and conservation of natural regeneration of different tree species. In Ghana, many floral biodiversity studies have been conducted at other moist semi-deciduous forests namely; Tinte Bepo Forest Reserve [9,23], Pra-Anum Forest Reserve [24] and Asenanyo River Forest [25] as well as other forest types like the moist evergreen forest type [23], including other forest reserves such as Atewa Range Forest Reserve [26], Anhwiaso North Forest Reserve [27] and Kakum Conservation Area [28]. Unfortunately, Bia Tano Forest Reserve (Bia Tano FR) has received very little floral biodiversity research assessment, despite being one of the most cherished forest reserves in Ghana because of its immense economic contribution to the forestry industry in the country. Against this background, the study sought to address the following objectives: (1) to examine regeneration dynamics between natural regeneration tree species with contradictory shade tolerance mechanism at undisturbed, slightly disturbed and disturbed areas within Bia Tano FR (2) to evaluate composition and diversity of natural regeneration tree species in different gaps and (3) to assess variabilities from the impact of varying intensities of anthropogenic disturbances on the conservation status of tree species in gap regeneration.

2. Materials and Methods

2.1. Study Site Description

Bia Tano FR is an African tropical moist semi-deciduous forest (latitude 6°53′–7°05′ N and longitude 2°32′–2°42′; 213 m–274 m a.s.l) (Figure 1) in the Ahafo Region of Ghana, West Africa. Bia Tano FR was established in 1937. It covers a total area of 18,197 ha and is being managed by Goaso Forest District under the auspices of Forest Services Division of Forestry Commission of Ghana.

Bia Tano FR shares boundaries with Goa Shelterbelt in North East, Bia Shelterbelt in North West and Bokoni Forest Reserve in the South. Geologically, Bia Tano FR is lying on metamorphosed sediments of lower Birimian age, predominantly schist and phyllites that are spread almost throughout the reserve's bedrock giving rise to weathered soils with low-activity clays. Ochrosol is the prevailing soil type. Annual precipitation ranges between 1250 and 1750 mm while the average monthly minimum and maximum temperatures are 26 °C and 29 °C correspondingly [29]. Great floristic diversity is the biodiversity hallmark of this reserve known for more than 190 plant species representing a manifest diversity of trees, shrubs, herbs and climbers or lianas. Bia Tano FR harbors abundant share of numerous common Ghanaian native tree species that normally find their optimum regeneration conditions in the moist semi-deciduous forest zone reaching a height between 50 to 60 m at the upper canopy creating two or more main cohorts at the forest canopy structure. Three International Union for Conservation of Nature Red List of Threatened Species™ (International Union for Conservation of Nature (IUCN) Red List) including *Baphia nitida, Celtis mildbraedii* and *Nesogordonia papaverifera* tree species are prevalent in this reserve.

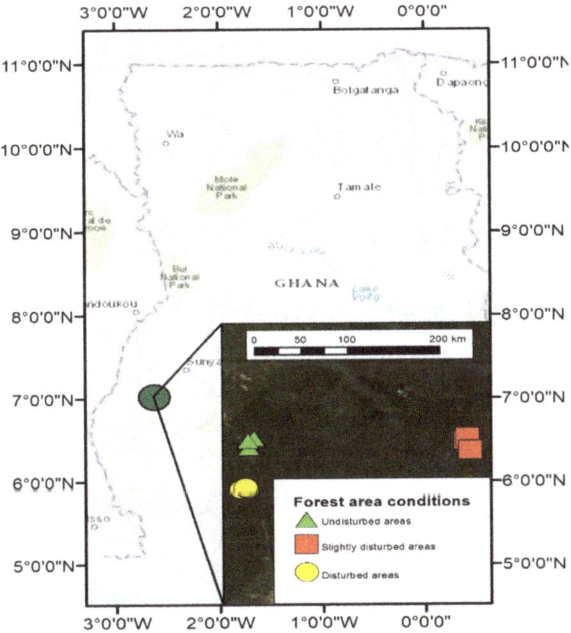

Figure 1. Locations of study areas within Bia Tano Forest Reserve of Ghana.

2.2. Sampling Design

Due to the rich floristic composition and diversity of tree species in Bia Tano FR, anthropogenic disturbances such as logging of timber trees (legal and/or illegal logging) and harvesting of Non-Timber Forest Products (NTFPs) are quite common. Nonetheless, logging stands tall and remains the most significant disturbance variable that explains variations in the various studied forest areas. Based on this, the study site was purposively stratified into three different forest areas to denote varying degrees of anthropogenic disturbances taking place in the forest reserve and assigning Forest Condition Score for tropical forests by Hawthorne and Abu-Juam [30] to illustrate the concise state of the various categorized forest site conditions. Therefore, the undisturbed area belongs to Forest Condition Score 1: this area shows rare signs (≤2%) of anthropogenic disturbance (logging) with a good canopy cover; slightly disturbed belongs to Forest Condition Score 2: this area shows few signs (>2 <10%) of anthropogenic disturbance with ecologically tolerant mosaic through good regeneration of timber trees and other forest plants, and finally, the disturbed area belongs to either Forest Condition Score 3 or 4: this area shows serious signs ($\geq 11 \leq 75\%$) of visible evidence of anthropogenic disturbance indicating slight to heavy disturbed canopy cover from unregulated logging operations.

To capture the proper appraisal of composition and diversity of natural regeneration tree species in gaps, which are the main focus of this study, i.e., gap regeneration at undisturbed, slightly disturbed and disturbed areas were considered. In the three forest areas, nine (9) gaps were randomly selected comprising three (3) each from undisturbed (Gap 1 = 1895 m^2; Gap 2 = 1244 m^2, Gap 3 = 1249 m^2), slightly disturbed (Gap 4 = 1425 m^2; Gap 5 = 1465 m^2, Gap 6 = 2005 m^2) and disturbed (Gap 7 = 286 m^2; Gap 8 = 607 m^2, Gap 9 = 884 m^2) areas, respectively. In each of the nine (9) selected gaps, a 20 m long four (4) distinct transects to the North (N), South (S), East (E) and West (W) directions were laid from gap centers. Then, all gap centers serving as reference points were georeferenced with the Geographical Positioning System gadget (Garmin GPSMAP, model 66st). After that, 10 circular subsampling regions of 1 m^2 area (56 cm radius) were marked at 2 m intervals along each transect (N-S-E-W). However, before this gap layout, one subsampling region was earlier created at all gap centers. A total of three hundred and sixty-nine (369) subsampling regions, comprising one hundred and twenty-three (123) (i.e., forty-one (41) per each gap) each for the undisturbed, slightly disturbed and disturbed areas, were used in evaluating natural regeneration tree species in gaps at Bia Tano FR.

2.3. Data Collection and Organization

Within every 1 m^2 subsampling region, natural regeneration tree species were first identified to species level, assessed morphologically and enumerated with the assistance of a botanist from the Department of Forest Science, the University of Energy and Natural Resources of Ghana and Photo Guide for Forest Trees in Ghana [31]. Species with the diameter at breast height (DBH) <10 cm and height ≤ 300 cm were considered as viable gap regeneration data. The DBH and height features of various natural regeneration tree species were measured using digital calipers (KENDO brand; 35,301 Model; ± 0.01 mm accuracy) and diameter tape (Stanley PowerLock brand), respectively. The data from the three studied forest areas were recorded on separate data sheets to prevent the mixing up of data.

Furthermore, gap regeneration was grouped into three functional tree species according to their natural shade tolerance mechanism (STM) in gap ecology namely; (1) Pioneers, i.e., sun-loving and ecologically light tolerable tree species; (2) nonpioneer light-demanding species (NPLD), i.e., intermediate ecologically light-shade tolerable tree species and, (3) shade bearers, i.e., ecologically shade-tolerant tree species [32,33]. To gain a better understanding of the effect of STM on the dynamics and distribution of tree species under gap regeneration is paramount in ecological studies. Therefore, we hypothesized that different growth behavior and pattern of tree species in gaps would occur as the reason for species' unique individual life-long ecological growth attributes. Based on the findings of previous studies that, in most tropical forests, there is a sequential ecological replacement of tree species essentially through regeneration shifts in species composition [12,22]. Most significant for pioneers, whose abundance begin to decline some years after logging (e.g., after 4 years [14], 7 years [24], 12 years [34], 20 years [35]) due to shorter life span from a more rapid growth character and higher mortality rates and consequently, at the later stages of succession, they are progressively replaced by a more slowly growing nonpioneer species [13].

2.4. Data Analysis

Paleontological Statistics (PAST) 3.24 educational software package [36] was utilized in calculating various assessed biodiversity indices. ANOVA and post hoc Tukey HSD analyses were carried out to show the significance of plant diversity variables means at $p < 0.05$ level, and all these analytical data procedures were executed with STATISTICA software package (13.4.0.14 version; TIBCO Corporation

Inc; Palo Alto, CA, USA). Microsoft Excel software package (2016 version; Redmond, WA, USA) was also used for the graphical presentations of results.

Additionally, relative density (%) was also calculated for each presented species.

$$\text{Relative density } = \frac{\text{Sum of particular species}}{\textit{Sum of all presented species}} \times 100 \tag{1}$$

Formulas for different analyzed biodiversity indices based on adopted Equations (Eqn.) from (2) to (7) according to Harper [37] are as follows.

Taxa (S) describes the number of presented species.

Total number (n) of presented individual species.

Simpson's index (1-D) measures the evenness of the natural regeneration community with its indicative values ranging from 0 to 1.

$$D = \sum_i \left(\frac{ni}{n}\right)^2 \tag{2}$$

Shannon diversity index (H) considered the number of individuals as well as the number of taxa varying from 0 for communities with only a single taxon to high values for communities with many taxa, each with few individuals.

$$H = -\sum_i \frac{ni}{n} In \frac{ni}{n} \tag{3}$$

Evenness (e^H/S) is another index that measures the evenness of individuals within the taxa community and is widely used to characterize species diversity in a community.

Menhinick's index (Mk).

$$\frac{S}{\sqrt{n}} \tag{4}$$

Margalef's index (Mf)

$$(S - 1)/In(n) \tag{5}$$

Equitability (J) measures the evenness whereby individuals are divided among the taxa present and is also known as Pielou's evenness.

$$J = \frac{H}{In(S)} \tag{6}$$

Fisher's alpha (α)

$$S = a*In(1 + n/a) \tag{7}$$

Meanwhile, two coefficient similarity indices were explicitly defined and calculated according to the equation formulated by Raup et al. [38] based on the binary function of absence-presence data.

Sorensen's coefficient similarity index (SCSI) emphasizes joint appearances of species present in both compared paired groups rather than their mismatches of species occurrences. This index is also termed as the Dice similarity coefficient [37]. SCSI was used for the pairwise comparison of species composition of the three different conditions of forest areas [27].

$$SCSI = 2M/(2M + N) \tag{8}$$

Jaccard coefficient similarity index (JCSI) is another form of similarity index for binary data which has equal relevance as SCSI [38]; however, it was used to give a clear state of the presence or absence of species between different paired forest areas [39].

$$JCSI = M/(M + N) \tag{9}$$

where; S is the number of taxa, n is the number of individuals, D is Dominance, ni is the number of individuals of taxon ith, In is the logarithm, a is Fisher's alpha, M is the number of species matches within the comparison pair and N is the sum number of species frequencies at forest areas in a column with a presence in just one row of species frequency.

3. Results

3.1. Description of Tree Species Composition in Gaps

A total of 752 individuals belonging to 63 species from 21 families and 52 genera were enumerated in the study (Table 1). Malvaceae and Meliaceae, with ten (10) species each, were the most diverse families, followed by Fabaceae with eight (8) species, and Moraceae with six (6) species while Melastomataceae, Ochnaceae, Olacaceae, Pandaceae, Putranjivaceae, Rutaceae, Sapindaceae and Urticaceae families were presented by only one species each. NPLD topped species composition with 27 species, followed by pioneers with 21 species while SB accrued the lowest count of 15 species. With conservation status of tree species, 32, 12, 12 and 7 species were acknowledged as Green Star, Pink Star, Scarlet Star and Red Star, respectively, under Conservation Star Ratings (CSR) system while under International Union for Conservation of Nature Red List of Threatened Species™ (IUCN Red List) system, higher count of Least Concern (LC) (34) species compared to Vulnerable (VU) (14) and Near Threatened (NT) (3) species were recognized alongside 12 other Not Evaluated (*NE*) not enlisted Red List species (NE*). *Nesogordonia papaverifera*, *Baphia nitida* and *Blighia sapida* species were commonly present in all gaps with their abundant proliferation records in Gap 9 (38%), Gap 4 (37%) and Gap 6 (13%), respectively.

Table 1. Composition of natural regeneration tree species in different gaps at Bia Tano Forest Reserve of Ghana. The abundance of individual species within each gap is presented by their relative densities. Columns show species names with authors, including their assigned taxonomic families [40]. Shade tolerance mechanism (STM) of species including Pi—Pioneers, NP_D—Nonpioneer light-demanding species and SB—Shade bearers as well as their conservation status in Ghana by standardized Conservation Star Ratings (CSR) showing Green Star (species with no particular conservation priority), Pink Star (species with low particular conservation priority), Red Star (species with some conservation priority) and Scarlet Star (species with high conservation priority) (order of increasing conservation concern) [30] and globally by LC—Least Concern species, NT—Near Threatened species and VU—Vulnerable species (increasing order of extinction risk) under International Union for Conservation of Nature Red List of Threatened Species™ (IUCN Red List) [41] alongside NE*—Not Evaluated (*NE*) not enlisted Red List species. Absence of species indicated by (—).

Species	Families	Conservation System			Relative Density (%)								
		STM	CSR	IUCN	Gap 1	Gap 2	Gap 3	Gap 4	Gap 5	Gap 6	Gap 7	Gap 8	Gap 9
Milicia excelsa (Welw.) C. C. Berg.	Moraceae	Pi	Scarlet	NT	—	—	—	—	—	—	—	0.94	—
Milicia regia (A.Chev.) C. C. Berg.	Moraceae	Pi	Scarlet	VU	—	—	—	—	—	1.92	—	—	—
Nauclea diderrichii (De Wild.) Merr.	Rubiaceae	Pi	Scarlet	VU	—	—	—	—	—	—	—	0.94	—
Triplochiton scleroxylon K. Schum.	Malvaceae	Pi	Scarlet	LC	—	—	2.13	1.41	2.13	—	—	0.94	1.12
Ceiba pentandra (L.) Gaertn	Malvaceae	Pi	Red	LC	0.81	—	—	—	—	—	0.59	1.89	—
Daniellia ogea (Harms) Holland	Fabaceae	Pi	Red	NT	2.44	10.64	21.28	1.41	—	—	2.94	4.72	2.25
Terminalia superba Engl. & Diels.	Combretaceae	Pi	Red	NE*	0.81	—	—	—	—	1.92	38.82	16.98	1.12
Antrocaryon micraster A. Chev. & Guillaumin.	Anacardiaceae	Pi	Pink	VU	—	—	—	—	—	—	—	0.94	—
Morus mesozygia Stapf.	Moraceae	Pi	Pink	NE*	0.81	—	—	—	2.13	—	—	0.94	—
Ricinodendron heudelotii (Baill.) Heckel.	Euphorbiaceae	Pi	Pink	NE*	0.81	—	—	—	2.13	—	—	—	—
Alstonia boonei De Wild.	Apocynaceae	Pi	Green	LC	—	—	—	—	—	—	—	0.94	—
Broussonetia papyrifera (L.) L'Hér. ex Vent.	Moraceae	Pi	Green	LC	3.25	—	—	—	—	—	0.59	0.94	1.12
Cleistopholis patens (Benth.) Engl. & Diels.	Annoraceae	Pi	Green	LC	—	—	—	—	4.26	—	—	—	—
Cola caricifolia (G.Don) K. Schum.	Malvaceae	Pi	Green	LC	—	—	2.13	1.41	—	—	—	—	—
Cola gigantea A. Chev.	Malvaceae	Pi	Green	LC	—	—	—	—	—	3.85	—	0.94	—
Drypetes gilgiana (Pax) Pax & K. Hoffm.	Putranjivaceae	Pi	Green	LC	0.81	—	—	—	—	—	—	—	—
Lannea welwitschii (Hiern) Engl.	Anacardiaceae	Pi	Green	LC	—	—	—	—	2.13	—	—	—	—
Musanga cecropioides R. Br. ex Tedlie.	Urticaceae	Pi	Green	LC	—	—	—	—	2.13	—	—	—	—
Psydrax subcordata (DC.) Bridson.	Rubiaceae	Pi	Green	NE*	1.63	—	—	—	—	—	—	—	—

Table 1. *Cont.*

Species	Families	Conservation System			Relative Density (%)								
		STM	CSR	IUCN	Gap 1	Gap 2	Gap 3	Gap 4	Gap 5	Gap 6	Gap 7	Gap 8	Gap 9
Tetrapleura tetraptera (Schum. & Thonn.) Taub.	Fabaceae	Pi	Green	LC	—	—	—	1.41	—	—	—	—	—
Zanthoxylum gilletii (De Wild.) P. G. Waterman.	Rutaceae	Pi	Green	LC	—	—	—	—	—	1.92	—	—	—
Aningeria robusta (A.Chev.) Aubrév. & Pellegr.	Sapotaceae	NPLD	Scarlet	LC	0.81	2.13	—	—	2.13	—	—	—	—
Entandrophragma candollei Harms.	Meliaceae	NPLD	Scarlet	VU	—	—	—	—	—	—	0.59	—	—
Entandrophragma cylindricum (Sprague) Sprague.	Meliaceae	NPLD	Scarlet	VU	0.81	—	—	—	—	—	—	—	—
Entandrophragma utile (Dawe & Sprague) Sprague.	Meliaceae	NPLD	Scarlet	VU	—	2.13	2.13	—	—	—	—	—	—
Guibourtia ehie (A.Chev.) J. Leonard.	Fabaceae	NPLD	Scarlet	LC	—	—	—	—	—	—	1.18	0.94	—
Khaya ivorensis A. Chev.	Meliaceae	NPLD	Scarlet	VU	—	—	—	—	—	1.92	—	—	—
Pterygota macrocarpa K. Schum.	Malvaceae	NPLD	Scarlet	VU	0.81	4.26	—	—	—	1.92	0.59	0.94	1.12
Amphimas pterocarpoides Harms.	Fabaceae	NPLD	Red	LC	0.81	—	—	—	—	—	—	—	—
Antiaris toxicaria Lesch.	Moraceae	NPLD	Red	LC	—	2.13	2.13	1.41	—	—	—	—	—
Mansonia altissima (A. Chev.) A. Chev.	Malvaceae	NPLD	Red	LC	3.25	—	—	4.23	2.13	—	7.06	16.98	1.12
Albizia zygia (DC.) J. F. Macbr.	Fabaceae	NPLD	Pink	LC	—	2.13	—	1.41	—	—	—	0.94	—
Celtis zenkeri Engl.	Cannabaceae	NPLD	Pink	LC	—	2.13	—	—	—	—	—	—	—
Funtumia elastica (Preuss) Stapf.	Apocynaceae	NPLD	Pink	LC	—	—	—	—	—	—	2.35	2.83	—
Sterculia oblonga Mast.	Malvaceae	NPLD	Pink	LC	—	—	—	4.23	—	—	—	0.94	—
Sterculia rhinopetala K. Schum.	Malvaceae	NPLD	Pink	LC	0.81	—	—	—	—	1.92	1.18	—	3.37
Terminalia ivorensis A. Chev.	Combretaceae	NPLD	Pink	VU	—	—	—	—	—	—	—	0.94	—
Albizia adianthifolia (Schum.) W. Wight.	Fabaceae	NPLD	Green	LC	—	4.26	—	—	—	—	—	—	—
Alchornea cordifolia (Schumach. & Thonn.) Müll. Arg.	Euphorbiaceae	NPLD	Green	LC	—	—	—	—	—	1.92	—	—	—
Blighia sapida K. D. Koenig.	Sapindaceae	NPLD	Green	LC	3.25	12.77	12.77	2.82	2.13	13.46	1.76	1.89	10.11
Campylospermum reticulatum Tiegh.	Ochnaceae	NPLD	Green	NE*	0.81	—	—	—	—	—	—	—	—
Duguetia staudtii (Engl. & Diels) Chatrou.	Annonaceae	NPLD	Green	LC	2.44	—	—	—	—	—	—	—	—
Garcinia kola Heckel.	Clusiaceae	NPLD	Green	VU	—	—	—	—	—	—	—	1.89	—
Morinda lucida A. Gray.	Rubiaceae	NPLD	Green	NE*	—	—	—	—	2.13	—	—	—	—
Trichilia lehmannii C. DC.	Meliaceae	NPLD	Green	NE*	—	—	—	4.23	—	—	—	—	1.12
Trichilia monadelpha (Thonn.) J. J. de Wilde.	Meliaceae	NPLD	Green	LC	2.44	—	—	—	—	—	—	—	—

Table 1. *Cont.*

Species	Families	Conservation System			Relative Density (%)								
		STM	CSR	IUCN	Gap 1	Gap 2	Gap 3	Gap 4	Gap 5	Gap 6	Gap 7	Gap 8	Gap 9
Trichilia tessmannii Harms.	Meliaceae	NPLD	Green	LC	—	—	—	—	2.13	3.85	—	—	—
Trilepisium madagascariense DC.	Moraceae	NPLD	Green	NE*	—	—	—	—	—	—	1.18	0.94	—
Khaya anthotheca (Welw.) C. DC.	Meliaceae	SB	Scarlet	VU	—	—	—	—	—	—	—	—	1.12
Guarea cedrata (A.Chev.) Pellegr.	Meliaceae	SB	Red	VU	0.81	—	—	—	—	—	—	—	—
Celtis mildbraedii Engl.	Cannabaceae	SB	Pink	LC	8.13	4.26	2.13	11.27	—	1.92	24.12	13.21	25.84
Nesogordonia papaverifera (A. Chev.) Capuron ex N. Hallé.	Malvaceae	SB	Pink	VU	30.89	19.15	10.64	21.13	19.15	9.62	4.71	11.32	38.20
Strombosia glaucescens Engl.	Olacaceae	SB	Pink	LC	—	—	—	—	4.26	13.46	2.94	3.77	—
Baphia nitida Lodd.	Fabaceae	SB	Green	LC	7.32	27.66	31.91	36.62	21.28	9.62	1.76	1.89	2.25
Carapa procera DC.	Meliaceae	SB	Green	LC	—	—	—	—	2.13	—	—	—	—
Chrysophyllum albidum G. Don.	Sapotaceae	SB	Green	NT	1.63	—	—	—	—	—	5.29	5.66	3.37
Cleidion gabonicum Baill.	Euphorbiaceae	SB	Green	NE*	8.94	—	—	—	—	—	—	—	—
Glyphaea brevis (Spreng.) Monach.	Malvaceae	SB	Green	NE*	0.81	2.13	—	—	—	—	—	—	—
Hymenostegia afzelii (Oliv.) Harms.	Fabaceae	SB	Green	NE*	—	—	—	—	17.02	3.85	—	—	—
Mallotus oppositifolius (Geiseler) Müll.Arg.	Euphorbiaceae	SB	Green	VU	—	2.13	10.64	—	—	—	—	—	—
Memecylon lateriflorum (G. Don) Bremek.	Melastomataceae	SB	Green	LC	—	—	—	—	6.38	17.31	—	1.89	6.74
Microdesmis keayana J. Léonard.	Pandaceae	SB	Green	NE*	13.82	2.13	2.13	7.04	—	5.77	2.35	1.89	—
Monodora myristica (Gaertn.) Dunal.	Annonaceae	SB	Green	LC	—	—	—	—	4.26	3.85	—	—	—
Total					100	100	100	100	100	100	100	100	100

Source: authors construct [2020].

3.2. Regeneration Dynamics between Pioneers and Shade Bearers in Gaps at Different Site Conditions

Our analysis investigating the ecological regeneration shift of pioneers to shade bearers in gap regeneration at three different forest areas in Figure 2 by logarithmic equations indicated that the rate of sequential replacement of pioneers by shade bearers in gaps at the slightly disturbed area (+1.2) was highly rapid compared to the undisturbed area (+7.8) whereas a worse ecological replacement was found at the disturbed area (+24.8). Besides, a stronger replacement relationship between concerned species within gaps at slightly disturbed (76%) and undisturbed (70%) areas were observed, while a much weaker ecological association was detected in disturbed areas (28%).

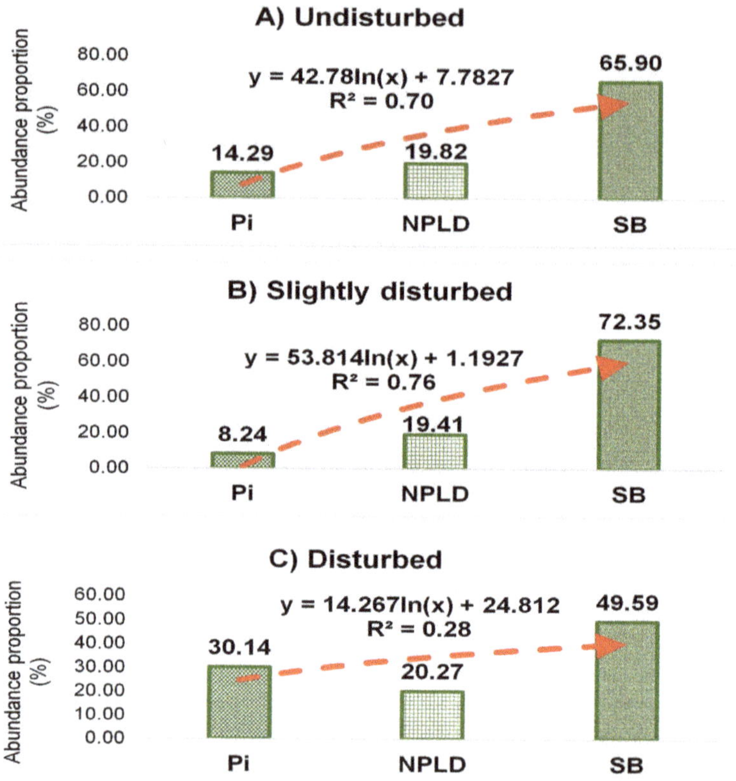

Figure 2. Ecological regeneration shift of tree species in gap regeneration under different forest areas (**A–C**). Pi—Pioneers, NPLD—Nonpioneer light-demanding species and SB—Shade bearers. Source: authors construct [2020].

3.3. Description of Tree Species Diversity in Gaps

Estimations of various engaged diversity indices (Taxa (mean range 11–28 S), Simpson's index (0.8–0.9 1-D), Shannon diversity (1.9–2.8 H), Evenness (0.4–0.7 e^H/S), Equitability (0.7–0.9 J), Menhinick (1.4–2.7 Mk), Margalef (2.5–5.7 Mf) and Fisher alpha (4.3–12.2 α)) revealed higher species diversity in gap regeneration (Table 2). Further, all estimated diversity indices showed significant biodiversity improvements in all studied gaps at $p < 0.05$ significance level. Yet still, there were no significant variations from gap to gap, except for Gap 8 (12 α) that significantly measured the highest species diversity for the Fisher alpha index (Table 2).

Table 2. Various biodiversity indices comprising Taxa (S), Individuals (N), Simpson's index (1-D), Shannon diversity index (H), Evenness index (e^H/S), Menhinick's richness (Mk), Margalef (Mf), Equitability (J) and Fisher alpha (α) presenting species diversity of gap regeneration at Bia Tano Forest Reserve of Ghana. Mean values with corresponding standard errors in parentheses are presented.

Gaps	S	N	Diversity Indices						
			1-D	H	e^H/S	Mk	Mf	J	α
Gap 1	26	123	0.86(0.02) ab	2.52(0.11) cd	0.49(0.05) ab	2.25(0.09) b	5.00(0.21) d	0.78(0.03) ab	9.49(0.58) b
Gap 2	15	47	0.85(0.03) ab	2.27(0.11) abcd	0.65(0.07) a	2.19(0.00) b	3.64(0.00) ac	0.84(0.04) ab	7.61(0.00) b
Gap 3	11	47	0.80(0.03) ab	1.92(0.12) a	0.64(0.07) ab	1.56(0.05) a	2.51(0.09) b	0.81(0.44) ab	4.31(0.21) a
Gap 4	14	71	0.79(0.04) ab	1.99(0.13) ab	0.54(0.06) ab	1.58 (0.08) a	2.89(0.16) ab	0.76(0.04) ab	4.86(0.36) a
Gap 5	18	47	0.87(0.02) ab	2.42(0.12) abcd	0.67(0.07) ab	2.48(0.15) bc	4.16(0.26) c	0.85(0.03) ab	9.65(1.02) b
Gap 6	18	52	0.89(0.02) ab	2.54(0.10) cd	0.73(0.06) b	2.40(0.09) bc	4.13(0.17) c	0.89(0.03) b	9.13(0.62) b
Gap 7	19	170	0.78(0.02) ab	2.02(0.10) abc	0.42(0.04) a	1.38(0.00) a	3.31(0.00) a	0.70(0.03) a	5.09(0.00) a
Gap 8	28	106	0.91(0.01) b	2.77(0.09) d	0.58(0.05) ab	2.69(0.03) c	5.72(0.07) d	0.83(0.03) ab	12.18(0.24) c
Gap 9	15	89	0.77(0.03) a	1.88(0.12) a	0.47(0.05) ab	1.48(0.11) a	2.89(0.22) ab	0.71(0.04) ab	4.69(0.48) a
	df		8	8	8	8	8	8	8
	F		3.97	8.49	3.13	37.97	44.35	3.21	32.51
	p		**	***	*	***	***	*	***

Means bearing same letters describe homogenous groups while those with different letters show significant differences at considered $p < 0.05$ (*), $p < 0.01$ (**) and $p < 0.001$ (***) significance levels, respectively. Source: authors construct [2020].

In addition, composite tree species in gaps for paired undisturbed × slightly disturbed areas as well as paired undisturbed × disturbed areas showed lower SCSI values of less than 0.5 representing 33% and 32% of mutual similar species, respectively, following JCSI measure. Nonetheless, paired slightly disturbed × disturbed areas showed a greater SCSI value of more than 0.5 representing a higher shared similar species of 37% by JCSI measure (Table 3).

Table 3. Using Sorensen's Coefficient Similarity Index (SCSI) and (JCSI) Jaccard Coefficient Similarity Index to compare species composition among three contrasting forest areas.

Paired Areas	Species Presence		Similarity Indices	
	Unique Species	Shared Species	SCSI	JCSI
Undisturbed × slightly disturbed	36	16	0.49	0.33
Undisturbed × disturbed	34	16	0.48	0.32
Slightly disturbed × disturbed	32	18	0.54	0.37

Source: authors construct [2020].

3.4. Description of Conservation Status of Tree Species in Gaps

Among species variations within site-condition level (vertical comparison) (Table 4), significant differences between species at undisturbed (df = 3, F = 6.136, $p = 0.018$) and slightly disturbed (df = 3, F = 16.542, $p = 0.001$) areas for CSR system and only slightly disturbed area (df = 3, F = 14.353, $p = 0.001$) for the IUCN Red List system were measured, respectively. Nevertheless, no significant difference ($p > 0.05$) was estimated among species at the disturbed area for the CSR system and repeatedly, at undisturbed and disturbed areas, respectively, for the IUCN Red List system. More so, among site-condition variations within species level (horizontal comparison), only VU species showed significant difference among studied site conditions (df = 2, F = 20.448, $p = 0.002$), revealing species higher significant mean regeneration density at disturbed areas compared to the other two forest areas (Table 4).

Table 4. Multiple comparison tests of the conservation status of tree species in gap regeneration under three different forest areas from a regional Conservation Star Ratings (CSR) and global International Union for Conservation of Nature (IUCN) Red List of Threatened Species™ perspectives. Mean values with corresponding standard errors in parentheses are presented.

Conservation System	Conservation Status	Mean Regeneration Density Per Hectare (n/ha)		
		Undisturbed	Slightly Disturbed	Disturbed
CSR	Green Star	244 (36.38) a	216 (28.11) a	449 (161.49)
	Pink Star	142 (68.64) ab	116 (37.15) ab	1134 (482.26)
	Red Star	65 (11.86) b	16 (9.75) b	1230 (874.59)
	Scarlet Star	21 (5.36) b	12 (2.45) b	85 (30.77)
IUCN Red List	LC	215 (8.00)	235 (47.68) a	1332 (636.5)
	NT	49 (15.92)	2 (2.34) b	248 (127.38)
	VU	138 (42.22) A	69 (19.24) b A	351 (31.99) B
	NE*	71 (59.23)	54 (13.10) b	967 (781.13)

Means bearing the same letters describe homogenous groups while those with different letters show significant differences at $p < 0.05$ significance level in columns (between species variation—lower-case letters) and rows (between forest areas variation—upper-case letters), respectively. However, means without any alphabet showed no significant differences at $p < 0.05$ significance level. LC—Least Concern, NT—Near Threatened, VU—Vulnerable and NE*—Not Evaluated (*NE*) not enlisted Red List species. Source: authors construct [2020].

4. Discussion

4.1. Assessment of Regeneration Dynamics between Pioneers and Shade Bearers in Gap Regeneration

The lower abundance proportion of pioneers in gap regeneration in this study substantiates that the proliferation of pioneers is temporary, and with time, regeneration composition gradually begins to shift to shade bearers [13,24]. The imbalance between high establishment and low survival rates of pioneers in gaps explains the observed regeneration dynamics of pioneers in Figure 2. This ecological behavior of pioneers creates an opportunity for nonpioneers to always take over regeneration under gap ecology by increasing their abundance proportions [13]. Another reason was the height of the bordering trees of gaps which transformed the incoming high direct light factor into a low direct light and high diffuse light conditions at gap microsites leading to the creations of unexpected comparable shading conditions, pronouncedly found after the fifth positions of subsampling regions (within 12–20 m^2 area). This limiting light conditions within gaps became an inconvenient growing condition for the development of pioneers but favorable for the unmatched proliferation of shade bearers. Previous studies by Čater et al. [15] proved our observation. Thus, it confirms that shade bearers have physiological and morphological plasticity, which grants them the adaptation ability to grow well and survive better under light environments like that of gaps [42]. Additionally, physiological processes such as photosynthesis could probably be linked to the observable fast replacement rate of pioneers to shade bearers within gaps at undisturbed and slightly disturbed areas. For instance, pioneers have high dark respiration, compensation point, saturation limit and quantum yield efficiency that only make them significantly flexible for growth under different light conditions, but not sufficient enough for their survival under total forest shade condition while nonpioneer species could illustrate relatively little increase in growth when light factor intensifies because of their low dark respiration, compensation and saturation points characteristics [23]. This explains why weighted abundance percentages of shade bearers (66–72%) in gaps at undisturbed and slightly disturbed areas were comparatively higher than those weighed for pioneers (8–14%), respectively. Similarly, this regeneration dynamics trend of ecological relationship between tree species with divergent STM in gap regeneration has been reported in the Brazilian rainforest [43]. Furthermore, a contradictory observation was made at the disturbed areas where regeneration shifts from pioneers to shade bearers composition in gaps was extremely slow and unpredictable. This gives an indication that disturbed areas probably created resilient grounds for the survivability of pioneers' regeneration as compared to other forest areas. The occurrence of the rate of pioneer regeneration reaching equilibrium with pioneer mortality rate as the result of regular adequate light supply and minimum soil surface disturbances from frequent anthropogenic disturbances in gaps at the disturbed area significantly defend this regeneration succession. Similar findings have been reported in tropical South American forests [13,44]. Hence, our result attests that pioneers are the worst variants, while shade bearers are the late opportunistic tree species in gap regeneration.

Generally, our results were contrary to studies of Abiem et al. [33] who found no apparent association between tree species with contrasting shade tolerance mechanisms.

4.2. Assessment of Composition and Diversity of Tree Species in Gap Regeneration

4.2.1. Species Composition

The higher species count (63) in this study (Table 1) justifies that tropical African forests are plant species enriched [4,33]. Our result is consistent with studies from China [21] and Brazil [13,44] but opposes to other works conducted in Europe, e.g., [15,19,45] and elsewhere in Asia, e.g., [20,46], where usually less than ten (10) tree species were enumerated in gap regeneration experiments in broadleaved mixed forests.

Additionally, the predominant composition of NPLD species in the total species composition was discovered in results. This finding has been reported also in mixed-oak forests at the southern Appalachians in America, where shade-intermediate *Quercus rubra* and *Acer rubrum* species abundantly occurred in the different studied gap plots [47]. It is self-evident that NPLD species sidelined cohabiting gap-dependent pioneers in gap regeneration. This finding corroborates an assertion that pioneers always experience increasing competition from nonpioneers [48]. The inherent light-shade character of NPLD species assisted them to adapt immediately and easily to the ever-changing light conditions within gaps and possibly gave them a leading advantage over other species groups in gaps. Moreover, higher enumeration of pioneers compared to shade bearers was noted in this study. The biological light-loving characteristic of pioneers propelled their flourishing regeneration performances in gaps. This observation supports the statement that pioneers prefer gaps/light environments/open areas for germination and growth [13,20,33]. Amongst, the STM groups of species, shade bearers (15 species) obtained the least representation in the overall species composition. The biological shade-tolerant character and ecological shade demanding behavior limited their regeneration capability in gaps. The combination of Pink Star, Red Star and Scarlet Star is termed as Reddish Stars [30]. It was detected that studied gaps presented balanced proportion (1:1) between Green Star and Reddish Stars in assessed species composition under CSR system while under IUCN Red List system, the proportion of LC species was comparatively higher than NT, VU species as well as the NE* species (11:1:4:4), respectively. The imbalance between not threatened LC and threatened NT, VU species in the IUCN Red List is comparable to the account of Pelletier et al. [49]. In forest practice, species proportion is the most frequently used applicable species composition indicator which describes how species occupy a particular growing space at stand level and also serves as a reliable growth and yield predictor at the forest level [50]. The comparable presence of Green Star (32 species) and Reddish Stars (31 species), as well as a higher proportion of LC species, suggests that Bia Tano FR has a good forest standing in terms of ecological resilience for promoting concurrent economic and ecological agenda of sustainable forest management in the long-term.

Notwithstanding, the assemblage of different tree species with varying biological, ecological and conservation characteristics in studied gaps was possibly due to the generation of a wide range of regeneration niches within gaps. These niches became ecologically stable microsites that provided optimal diverse light condition requirements for different species with distinctive ecological growth demands as well as different developmental life history attributes. Our findings that gaps are excellent assemblage places for different species agree with a couple of studies (e.g., [16,20]) but protest strongly to an assertion from Jaloviar et al. [19] that gaps do not necessarily provide primary environments for regeneration and species coexistence because they found the organization of relatively homogenous tree species composition in gaps during their studies.

4.2.2. Species Diversity and Similarity Check

Many studies have recounted the significance of species diversity on forest functioning, productivity, stability and provision of ecosystem services [45,50,51]. However, this study would be contributing to the important effect of gap regeneration on tree species diversity from the tropical African viewpoint. From the study results, a wealthy worth of tree species was expressly observed (Table 1), which indicated rich biodiversity across all evaluated gaps in Bia Tano FR. Presented high species diversity and richness in gaps (Table 2) could primarily be linked to two explanatory factors. The first explanatory factor being that gaps as the growing sites for the investigation of tree species diversity. Therefore, gaps as growing spaces usually offer adequate amounts of needful growth resources for seed germination and seedling development due to their ability to sequestrate resources [42]. The influence of gaps on microenvironmental heterogeneity factors such as variability in light, soil moisture, soil temperature and nutrient availability leads to the promotion and maintenance of species diversity within gap sites [52]. The second explanatory factor being that Bia Tano FR as a growing habitat for the investigation of tree species diversity. Therefore, the ecological settings within this particular tropical moist semi-deciduous forest type provided a broad spectrum of optimal growing niches for different suites of tree species [53]. This, in effect, significantly influenced and improved the organization of high species diversity and richness within gaps across the three studied forest areas. This validates the statement that Bia Tano FR has a magnificent species diversity comprising more than 190 plant species serving as one of the outstanding floral biodiversity repositories in Ghana [53]. Further, it was observed that gaps (i.e., Gaps 7–9) at disturbed areas enumerated the highest species individuals in the study. The massive contribution of native pioneers' regeneration significantly explained this observation. Constant high light availability in gaps [23], regular forest disturbances including other recurrent anthropogenic activities and fast-spreading nature of pioneers in tropical forests [31] seemingly influenced the prolific natural regeneration behavior of pioneers at disturbed areas. However, copious seed supply from parent stands, seed dispersal from the canopy cover, seeds repository and available healthy mother trees at undisturbed areas influenced the high number of species individuals in Gap 1. A similar observation was made in the tropical rainforest of the Congolese Basin in the Republic of Congo, where Ifo et al. [39] detected that studied plots at degraded and primary forests achieved the highest taxa and species individuals' records and they affirmatively associated soil type, rainfall trends, anthropogenic action, and land-use change as the underlying reasons for their observation. Contrary to our results, the range of values for Simpson's (0.96–0.97 1-D), Shannon diversity (3.77–3.90 H), Menhinick (2.51–2.82 Mk), Margalef (11.62–13.06 Mf), and Fisher alpha (20.72–24.16 α) indices obtained in another study at Kakum Conservation Area (highly protected area in Ghana) by Wiafe [28] were comparatively higher than the respective range of values estimated in this study. Nonetheless, the maximum attained value of the Equitability index (0.89 J) was relatively higher than the range of values for the same diversity index in Wiafe's [28] study.

According to the literature [27], if the SCSI value is lower than 0.5, then the paired communities share different species composition, but if the index is greater than 0.5, then the paired communities share similar species composition. Therefore, the species similarity tests (Table 3) revealed a clear distinction of species composition in gaps between undisturbed × slightly disturbed areas and again, between undisturbed × disturbed areas, respectively. Nonetheless, it could be deduced that slightly disturbed × disturbed areas shared similar species composition. Apart from paired undisturbed × disturbed areas (0.48) that shared parallel outcome with results of undisturbed and heavily disturbed areas (0.33) in Wiafe [25], both undisturbed × slightly disturbed (0.49) and slightly disturbed × disturbed (0.54) areas shared contradictory results with their counterparts; undisturbed and slightly disturbed (0.55) and slightly disturbed and heavily disturbed (0.47) areas, respectively, in the same mentioned study. Additionally, it was observed that gaps at undisturbed × slightly disturbed areas shared a higher number of different tree species, followed by undisturbed × disturbed areas while slightly disturbed × disturbed areas presented poor species dissimilarity records (Table 3). The dissimilarities of unique species composition could probably be associated to the contrasting ecological settings from different intensities of forest disturbance events at the studied forest areas, differences in microenvironmental conditions at the stand level together with differential conditions within gap microsites that substantively determine species germination, establishment and development at the forest understory. Conversely, similar forest site conditions due to the slight variations of forest disturbance regimes (intensity, scale, and frequency) between slightly disturbed and disturbed areas created regeneration avenues that favored tree species with similar life history attributes ending up in the assemblage of a high count of common species (18 species).

4.3. Assessment of Conservation Status of Tree Species in Gap Regeneration

Species conservation schemes have become an integral component of biodiversity and conservation studies of forests because they give a clear picture of the conservation status of various tree species and at the same time, bring out the conservation potential of a forest by drawing attention to the exact state of tree species vulnerability to life-long exploitation attributes and intensity alongside revealing the current state of forest degradation in general. For this paper, the considered CSR and IUCN Red List systems are particularly significant in evaluating the comprehensive conservation status of all identified tree species in gap regeneration. Therefore, the CSR system is a robust scientific-based accepted species conservation scheme for assessing forest trees in Ghana describing the rarest to common tree species by *Star Ratings* [30] while the IUCN Red List system is a key tool for conservation [49] and authoritative guide to the status of biodiversity relevant to all species (i.e., plants and animals) and all regions of the world [41] revealing the global level conservation status of tree species in this study.

From the presented CSR results (Table 4), four Stars out of the seven (7) (i.e., Green Star, Pink Star, Red Star, Scarlet Star, Blue Star, Gold Star and Black Star) recognized Stars under CSR were found in the study. The prevalent composition of Pink Star and Green Star species in gaps at undisturbed and slightly disturbed areas could be attributed to the dominance of these tree species in the forest ecosystem. Our results are consistent with the findings of Akoto et al. [27] who enumerated the abundance proportion of natural regeneration of Green Star species at three different vegetation types, respectively. Besides, these groups of species are not under any form of exploitation threats because they have moderate to low commercial interests [28]. Hence, these two Star Ratings have no particular conservation concerns for now in Ghana. Further, low seeds production coupled with poor quality and inviable seeds by few unproductive and overly matured Scarlet and Red Stars trees due to the restricted passive forest management guideline at undisturbed areas while overexploitation of mother Scarlet and Red Star trees at slightly disturbed areas could be explained for the unimpressive regeneration performances of both Star species within gaps at those forest areas. Generally, the low regeneration composition of Scarlet Star and Red Star species in gap regeneration was possible to challenge their status as commodities of high financial value and utilitarian merit by undermining their population structure and degrading their bioquality in the study site. However, no species belonging to Blue Star, Gold Star and Black Star (high conservation concern) were encountered across all studied forest areas. Similar studies in other forest reserves of the moist semi-deciduous forest zone of Ghana documented the same findings [54,55]. The rarity of these species in gap regeneration projects substantive evidence on forest degradation in terms of biological legacy and genetic quality at the study site. The values of the various Star Ratings could be helpful guides for the identification of genetic hotspots within forests [22]. In view of this, disturbed areas seem to be genetically viable for biodiversity hotspots of Green Star, Pink Star, Red Star, and Scarlet Star species of the study area.

For IUCN Red List system, the significant widely spread of LC species within gaps at slightly disturbed areas indicated the predominant spreading of their mother producing trees which required no protection priority because of minimal hunting concern by loggers while nature-degrading overexploitation regimes of various tree species for economic benefits were associated with the significant low regeneration of NT and VU species [56]. This finding supports the statement that reducing the intensity of logging is important for the conservation of species [1]. In addition, among the three forest areas variations, the disturbed area deemed fit for the protection and conservation of VU species as it significantly maintained abundant regeneration of these species. This finding is different from the assertion that rising deforestation might increase the likelihood that vulnerable species would go extinct [3]. However, our results indicated a good number (12 species, 19%) of NE* species. This has been mentioned in another study that only 5% of plant species housed in the Global Biodiversity Information Facility were presently listed on the IUCN Red List and as at now, the conservation status of most plant species is unknown [49]. Meanwhile, Walker et al. [57] addressed that plant species assessed for the IUCN Red List were not randomly chosen but reflected geographic and taxonomic preferences, including their relevance to list threatened species. Finally, there was no regeneration indication of higher extinction risk species (Endangered (EN), Critically Endangered (CR), Extinct in the wild (EW) and Extinct (E)). Our results agree with the claim that most tropical plant species lack extinction risk assessments, limiting scientists' ability to identify conservation priorities particularly in tropical African flora where potential threats of a high level of species extinction risk (33%) are quite imminent [2]. Generally, our results underlined the winning position of Bia Tano FR to championing the global call for sustainable management, protection and conservation of tree species [56,58,59] through excellent ecological forest restoration and recovery mechanisms within gaps after various degrees of forest disturbances. Therefore, the urgent need for forest managers to prioritize the protection of this forest reserve towards the enhancement of species biodiversity and conservation.

5. Conclusions

In a quest to improve the diversity and conservation of native tree species in tropical African forests, gap regeneration remains all-important nature-promoting silviculture practice and ecosystem-based strategy for attaining these ecological goals. Bia Tano FR is a very important forest reserve in Ghana due to its rich floral diversity and resourcefulness for future conservation efforts by providing suitable habitats for several threatened tree species. However, the sustainability of Bia Tano FR largely depends on species structure, composition, and diversity. In this study, gaps were proved as a crucial silviculture intervention and indispensable growing spaces in the forest that offered remarkable conditions for high natural regeneration yield and diversity. It was observed that gap creations positively affected the structural complexity of Bia Tano FR. Regimes of anthropogenic disturbances could not significantly explain species diversity but became the single measured variable that significantly explained species composition, regeneration shift and conservation variations. Shade bearers were abundant and stable at undisturbed and slightly disturbed areas, while pioneers were abundant and stable at the disturbed area. Additionally, shade bearers were the most rapidly replacing tree species while pioneers were the fast diminishing tree species in succession under gap regeneration. Besides, Green Star and Pink Star were the frequent occurring species while Red Star and Scarlet Star were scarcely occurring species at undisturbed and slightly disturbed areas, respectively, at the regional scale while at the global scale, LC species attained predominant records at the expense of NT and VU species significantly at the slightly disturbed area while among forest areas comparison test revealed disturbed area as potent regeneration grounds for the competitive advantage of VU species.

Overall, the results add to the body of evidence for the impact of different intensities of disturbances on the population structure of tropical plant diversity. Our novel conservation status appraisal approach, crucial to the in-depth understanding of African tropical tree diversity represents a significant step to data-driven comprehensive conservation assessments applicable beyond the continental scales. Presented detailed analyses of results plus robust findings would enable conservation researchers and managers to identify taxonomically verified and high-quality species distribution datasets including unassessed species that are not currently recognized as species of global concern. Therefore, our study is important for the identification of conservation hotspots in need of further evaluation for sustainable forest management prioritization. Additionally, our results provide conservation-relevant knowledge that promotes the management of floral biodiversity based on conservation guidelines, gap regeneration: a pressing strategy for tree species richness and preservation worldwide.

Lastly, we strongly discourage anthropogenic activities (i.e., overexploitation of species, unsustainable economic and habitat destruction activities) that erode biological legacies but encourage forest protection programs that safeguard the world's threatened tree species from extinction. Additionally, we recommend further studies comparing the results of this study with other tropical African forests within the moist semi-deciduous zone so as to validate the actual biodiversity and conservation stance of this forest type.

Author Contributions: Conceptualization, M.E.H.; data curation, M.E.H. and R.P.; formal analysis, M.E.H.; funding acquisition, M.E.H. and R.P.; investigation, M.E.H. and R.P.; methodology, R.P.; project administration, R.P.; software, M.E.H.; supervision, M.E.H. and R.P.; visualization, M.E.H.; writing—original draft, M.E.H.; writing—review and editing, M.E.H. and R.P. All authors have read and agreed to the published version of the manuscript.

Funding: This research was funded by the Internal Grant Agency of Mendel University in Brno (LDF_VP_2019015) and the Framework of Bilateral Mobility Program for Traineeship of Doctoral Students, MENDELU.

Acknowledgments: The authors are grateful to the Forest Services Division of the Forestry Commission of Ghana, Goaso Forest District, for giving approval for the work to be conducted in the Bia Tano Forest Reserve and for granting personnel and field assistance. We are also appreciative to the Department of Forest Science, University of Energy and Natural Resources of Ghana for helping with the identification of the plant species. We are also thankful to Augustine Gyedu of the Forest Services Division at Kumasi, William Dumenu, Shalom Addo-Danso and Akwasi Duah Gyamfi of the Forest Research Institute of Ghana for their valuable contributions and consultations during traineeship. Additionally, many thanks go to the Department of Silviculture, Mendel University in Brno for the immeasurable technical and logistics support. Not forgetting, Zdeněk Patočka from the Department of Forest Management and Applied Geoinformatics, Mendel University in Brno for the production of the study site map.

Conflicts of Interest: The authors declare no conflicts of interest.

References

1. Burivalova, Z.; Allnutt, T.F.; Rademacher, D.; Schlemm, A.; Wilcove, D.S.; Butler, R.A. What works in tropical forest conservation, and what does not: Effectiveness of four strategies in terms of environmental, social, and economic outcomes. *Conserv. Sci. Pract.* **2019**, *1*, e28. [CrossRef]

2. Stévart, T.; Dauby, G.; Lowry, P.P.; Blach-Overgaard, A.; Droissart, V.; Harris, D.J.; Mackinder, B.A.; Schatz, G.E.; Sonké, B.; Sosef, M.S.M.; et al. A third of the tropical African flora is potentially threatened with extinction. *Sci. Adv.* **2019**, *5*, eaax9444. [CrossRef]

3. Senior, R.A.; Hill, J.K.; Edwards, D.P. Global loss of climate connectivity in tropical forests. *Nat. Clim. Chang.* **2019**, *9*, 623–626. [CrossRef]

4. Pirie, M.D. Remarkable insights into processes shaping African tropical tree diversity. *Peer Community Evol. Biol.* **2020**, *1*, 100094. [CrossRef]

5. Nair, C.T.S.; Tieguhong, J. *African Forests and Forestry: An Overview. A Report Prepared for the Project. Lessons Learnt on Sustainable Forest Management in Africa*; Royal Swedish Academy of Agriculture and Forestry: Stockholm, Sweden; African Forest Research Network (AFORNET) at the African Academy of Sciences (AAS): Nairobi, Kenya; Food and Agriculture Organisation of the United Nations: Rome, Italy, 2004.

6. Dutta, G.; Devi, A. Plant diversity, population structure, and regeneration status in disturbed tropical forests in Assam, northeast India. *J. For. Res.* **2013**, *24*, 715–720. [CrossRef]

7. Kacholi, D.S. Edge-Interior Disparities in Tree Species and Structural Composition of the Kilengwe Forest in Morogoro Region, Tanzania. *Int. Sch. Res. Not.* **2014**, *2014*, 1–8. [CrossRef]

8. Karsten, R.J.; Jovanovic, M.; Meilby, H.; Perales, E.; Reynel, C. Regeneration in canopy gaps of tierra-firme forest in the Peruvian Amazon: Comparing reduced impact logging and natural, unmanaged forests. *For. Ecol. Manag.* **2013**, *310*, 663–671. [CrossRef]

9. Lawer, E.A.; Baatuuwie, B.N.; Ochire-Boadu, K.; Asante, J.W. Preliminary assessment of the effects of anthropogenic activities on vegetation cover and natural regeneration in a moist semi-deciduous forest of Ghana. *Int. J. Ecosyst.* **2013**, *3*, 148–156.

10. Fischer, R.; Bohn, F.; De Paula, M.D.; Dislich, C.; Groeneveld, J.; Gutierrez, A.G.; Kazmierczak, M.; Knapp, N.; Lehmann, S.; Paulick, S.; et al. Lessons learned from applying a forest gap model to understand ecosystem and carbon dynamics of complex tropical forests. *Ecol. Model.* **2016**, *326*, 124–133. [CrossRef]

11. Kumar, A.; Marcot, B.G.; Saxena, A. Tree species diversity and distribution patterns in tropical forests of Garo Hills. *Curr. Sci.* **2006**, *91*, 1370–1381.

12. Rahman, H.; Khan, A.S.A.; Roy, B.; Fardusi, M.J. Assessment of natural regeneration status and diversity of tree species in the biodiversity conservation areas of Northeastern Bangladesh. *J. For. Res.* **2011**, *22*, 551–559. [CrossRef]

13. De Carvalho, A.L.; D'Oliveira, M.V.N.; Putz, F.E.; De Oliveira, L.C. Natural regeneration of trees in selectively logged forest in western Amazonia. *For. Ecol. Manag.* **2017**, *392*, 36–44. [CrossRef]

14. Park, A.; Justiniano, M.J.; Fredericksen, T.S. Natural regeneration and environmental relationships of tree species in logging gaps in a Bolivian tropical forest. *For. Ecol. Manag.* **2005**, *217*, 147–157. [CrossRef]

15. Čater, M.; Diaci, J.; Rozenbergar, D. Gap size and position influence variable response of Fagus sylvatica L. and Abies alba Mill. *For. Ecol. Manag.* **2014**, *325*, 128–135. [CrossRef]

16. Bentos, T.V.; Nascimento, H.E.; dos Anjos Vizcarra, M.; Williamson, G.B. Effects of light gaps and topography on Amazon secondary forest: Changes in species richness and community composition. *For. Ecol. Manag.* **2017**, *396*, 124–131. [CrossRef]

17. Yamamoto, S.-I. Forest gap dynamics and tree regeneration. *J. For. Res.* **2000**, *5*, 223–229. [CrossRef]

18. Brokaw, N.; Busing, R.T. Niche versus chance and tree diversity in forest gaps. *Trends Ecol. Evol.* **2000**, *15*, 183–188. [CrossRef]

19. Jaloviar, P.; Sedmáková, D.; Pittner, J.; Danková, L.J.; Kucbel, S.; Sedmák, R.; Saniga, M. Gap Structure and Regeneration in the Mixed Old-Growth Forests of National Nature Reserve Sitno, Slovakia. *Forests* **2020**, *11*, 81. [CrossRef]

20. Lu, D.; Zhang, G.; Jiao-Jun, Z.; Wang, G.G.; Zhu, C.; Yan, Q.; Zhang, J. Early natural regeneration patterns of woody species within gaps in a temperate secondary forest. *Eur. J. For. Res.* **2019**, *138*, 991–1003. [CrossRef]

21. Zhu, J.; Lu, D.; Zhang, W. Effects of gaps on regeneration of woody plants: A meta-analysis. *J. For. Res.* **2014**, *25*, 501–510. [CrossRef]

22. Agyarko, T. *Forestry Outlook Study for Africa (FOSA)*; Country Report; Ministry of Lands and Forestry: Accra, Ghana, 2001; pp. 15–22.

23. Agyeman, V.K.; Swaine, M.D.; Thompson, J.; Kyereh, B.; Duah-Gyamfi, A.; Foli, E.G.; Adu-Bredu, S. A comparison of tree seedling growth in artificial gaps of different sizes in two contrasting forest types. *Ghana J. For.* **2011**, *26*, 14–40. [CrossRef]

24. Duah-Gyamfi, A.; Kyereh, B.; Adam, K.A.; Agyeman, V.K.; Swaine, M.D. Natural regeneration dynamics of tree seedlings on skid trails and tree gaps following selective logging in a tropical moist semi-deciduous forest in Ghana. *Open For. Res.* **2014**, *41*, 49. [CrossRef]

25. Wiafe, E.D. Tree regeneration after logging in rain-forest ecosystem. *Res. J. Biol.* **2014**, *2*, 18–28.

26. Kusimi, J.M. Characterizing land disturbance in Atewa Range Forest Reserve and Buffer Zone. *Land Use Policy* **2015**, *49*, 471–482. [CrossRef]

27. Akoto, S.D.; Asare, A.; Gyabaa, G. Natural regeneration diversity and composition of native tree species under monoculture, mixed culture plantation and natural forest. *Int. Res. J. Nat. Sci.* **2015**, *3*, 24–38.

28. Wiafe, E.D. Tree Species Composition of Kakum Conservation Area in Ghana. *Ecol. Evol. Biol.* **2016**, *21*, 14.

29. Gelens, M.F.; van Leeuwen, L.M.; Hussin, Y.A. *Geo-Information Applications for Off-Reserve Tree Management, Ghana*; Tropenbos International: Wageningen, The Netherlands, 2010; p. 3.

30. Hawthorne, W.D.; Abu-Juam, M. *Forest Protection in Ghana with Particular Reference to Vegetation and Species*; IUCN Press: Cambridge, UK, 1995; pp. 12–27.

31. Hawthorne, W.D.; Gyakari, N. *Photoguide for the Forest Trees of Ghana: A Tree-Spotter's Field Guide for Identifying the Largest Trees*; Oxford Forestry Institute: Oxford, UK, 2006; pp. 18–348.

32. Hawthorne, W.D. *Ecological profiles of Ghanaian Forest Trees*; Tropical Forestry Paper; Oxford Forestry Institute: Oxford, UK, 1995; p. 29.

33. Abiem, I.; Arellano, G.; Kenfack, D.; Chapman, H. Afromontane Forest Diversity and the Role of Grassland-Forest Transition in Tree Species Distribution. *Diversity* **2020**, *12*, 30. [CrossRef]

34. Kuusipalo, J.; Jafarsidik, Y.; Ådjers, G.; Tuomela, K. Population dynamics of tree seedlings in a mixed dipterocarp rainforest before and after logging and crown liberation. *For. Ecol. Manag.* **1996**, *81*, 85–94. [CrossRef]

35. Dekker, M.; De Graaf, N. Pioneer and climax tree regeneration following selective logging with silviculture in Suriname. *For. Ecol. Manag.* **2003**, *172*, 183–190. [CrossRef]

36. Hammer, Ø.; Harper, D.A.; Ryan, P.D. PAST: Paleontological statistics software package for education and data analysis. *Palaeontol. Electron.* **2001**, *4*, 9.

37. Wagner, P.J.; Harper, D.A.T. Numerical Palaeobiology: Computer-Based Modelling and Analysis of Fossils and Their Distributions. *PALAIOS* **2000**, *15*, 364–366. [CrossRef]

38. Raup, D.M.; Crick, R.E. Measurement of faunal similarity in paleontology. *J. Paleontol.* **1979**, *53*, 1213–1227.

39. Ifo, S.A.; Moutsambote, J.M.; Koubouana, F.; Yoka, J.; Ndzai, S.F.; Bouetou-Kadilamio, L.N.O.; Mampouya, H.; Jourdain, C.; Bocko, Y.; Mantota, A.B.; et al. Tree Species Diversity, Richness, and Similarity in Intact and Degraded Forest in the Tropical Rainforest of the Congo Basin: Case of the Forest of Likouala in the Republic of Congo. *Int. J. For. Res.* **2016**, *2016*, 1–12. [CrossRef]

40. World Flora Online. Available online: http://www.worldfloraonline.org/ (accessed on 15 July 2020).

41. International Union for Conservation of Nature. 2006 IUCN Red List of Threatened Species, 2020 (Version 2020-2). Available online: http://www.iucnredlist.org/ (accessed on 17 July 2020).

42. Muscolo, A.; Bagnato, S.; Sidari, M.; Mercurio, R. A review of the roles of forest canopy gaps. *J. For. Res.* **2014**, *25*, 725–736. [CrossRef]

43. Gomes, E.P.C.; Mantovani, W.; Kageyama, P.Y. Mortality and recruitment of trees in a secondary montane rain forest in southeastern Brazil. *Braz. J. Biol.* **2003**, *63*, 47–60. [CrossRef]

44. Darrigo, M.R.; Venticinque, E.; Dos Santos, F.A.M. Effects of reduced impact logging on the forest regeneration in the central Amazonia. *For. Ecol. Manag.* **2016**, *360*, 52–59. [CrossRef]

45. Pretzsch, H.; Forrester, D.I.; Rötzer, T. Representation of species mixing in forest growth models. A review and perspective. *Ecol. Model.* **2015**, *313*, 276–292. [CrossRef]

46. Pourbabaei, H.; Haddadi-Moghaddam, H.; Begyom-Faghir, M.A.; Abedi, T. The influence of gap size on plant species diversity and composition in beech (Fagus orientalis) forests, Ramsar, Mazandaran Province, North of Iran. *Biodivers. J. Biol. Divers.* **2013**, *14*, 2. [CrossRef]

47. Clinton, B.D.; Boring, L.R.; Swank, W.T. Regeneration Patterns in Canopy Gaps of Mixed-Oak Forests of the Southern Appalachians: Influences of Topographic Position and Evergreen Understory. *Am. Midl. Nat.* **1994**, *132*, 308–319. [CrossRef]

48. Swaine, M.D.; Agyeman, V.K. Enhanced Tree Recruitment Following Logging in Two Forest Reserves in Ghana. *Biotropica* **2008**, *40*, 370–374. [CrossRef]

49. Pelletier, T.A.; Carstens, B.C.; Tank, D.C.; Sullivan, J.; Espíndola, A. Predicting plant conservation priorities on a global scale. *Proc. Natl. Acad. Sci. USA* **2018**, *115*, 13027–13032. [CrossRef] [PubMed]

50. del Río, M.; Pretzsch, H.; Alberdi, I.; Bielak, K.; Bravo, F.; Brunner, A.; Condés, S.; Ducey, M.J.; Fonseca, T.; von Lüpke, N.; et al. *Characterization of Mixed Forests in Dynamics, Silviculture and Management of Mixed Forests*; Springer: Berlin/Heidelberg, Germany, 2018; pp. 27–71.

51. Liang, J.; Crowther, T.W.; Picard, N.; Wiser, S.; Zhou, M.; Alberti, G.; Schulze, E.-D.; McGuire, A.D.; Bozzato, F.; Pretzsch, H.; et al. Positive biodiversity-productivity relationship predominant in global forests. *Science* **2016**, *354*, aaf8957. [CrossRef] [PubMed]

52. Xu, J.; Lie, G.; Xue, L. Effects of gap size on diversity of soil fauna in a Cunninghamia lanceolata stand damaged by an ice storm in southern China. *J. For. Res.* **2016**, *27*, 1427–1434. [CrossRef]

53. *The State of the World's Forest Genetic Resources*; Country Report; Ministry of Lands and Natural Resources: Accra, Ghana, 2012; pp. 20–24.

54. Adusei, Y.Y. Assessment of the Natural Regeneration Potential of Reclaimed Mined Sites: Case Study of Owere Mines Limited. Master's Thesis, University of Energy and Natural Resources, Sunyani, Ghana, 12 May 2017.

55. Yalley, M.K.; Adusu, D.; Bunyamin, A.-R.; Okyere, I.; Asare, A. Natural Regeneration of Indigenous Tree Species in Broussonetia papyrifera Invaded Sites in Pra-Anum Forest Reserve. *Int. J. For. Res.* **2020**, *2020*, 1–9. [CrossRef]

56. Maljean-Dubois, S. The intergovernmental science-policy platform on biodiversity and ecosystem services (IPBES). *Int. J. Bioethique* **2014**, *25*, 55–73. [CrossRef]

57. Walker, B.E.; Leão, T.C.C.; Bachman, S.P.; Bolam, F.C.; Nic Lughadha, E. Caution Needed When Predicting Species Threat Status for Conservation Prioritization on a Global Scale. *Front. Plant Sci.* **2020**, *11*, 520. [CrossRef]
58. Convention on International Trade in Endangered Species of Wild Fauna and Flora. Available online: https://cites.org/eng/message_SG_Ivonne_Higuero_idb2020_22052020 (accessed on 21 July 2020).
59. Global Tree Assessment. Available online: https://www.globaltreeassessment.org/ (accessed on 21 July 2020).

Article

Spatial and Temporal Trends of Burnt Area in Angola: Implications for Natural Vegetation and Protected Area Management

Silvia Catarino [1,2,*], Maria Manuel Romeiras [1,3,*], Rui Figueira [1,4], Valentine Aubard [2], João M. N. Silva [2] and José M. C. Pereira [2]

[1] Linking Landscape, Environment, Agriculture and Food (LEAF), School of Agriculture, University of Lisbon, Tapada da Ajuda, 1349-017 Lisbon, Portugal; ruifigueira@isa.ulisboa.pt

[2] Forest Research Centre (CEF), School of Agriculture, University of Lisbon, Tapada da Ajuda, 1349-017 Lisbon, Portugal; vaubard@isa.ulisboa.pt (V.A.); joaosilva@isa.ulisboa.pt (J.M.N.S.); jmcpereira@isa.ulisboa.pt (J.M.C.P.)

[3] Centre for Ecology, Evolution and Environmental Changes (cE3c), Faculty of Sciences, University of Lisbon, 1749-016 Lisbon, Portugal

[4] Research Centre in Biodiversity and Genetic Resources (CIBIO/InBIO), School of Agriculture, University of Lisbon, Tapada da Ajuda, 1349-017 Lisbon, Portugal

* Correspondence: scatarino@isa.ulisboa.pt (S.C.); mmromeiras@isa.ulisboa.pt (M.M.R.)

Received: 18 June 2020; Accepted: 5 August 2020; Published: 9 August 2020

Abstract: Fire is a key driver of natural ecosystems in Africa. However, human activity and climate change have altered fire frequency and severity, with negative consequences for biodiversity conservation. Angola ranks among the countries with the highest fire activity in sub-Saharan Africa. In this study, we investigated the spatial and temporal trends of the annual burnt area in Angola, from 2001 to 2019, and their association with terrestrial ecoregions, land cover, and protected areas. Based on satellite imagery, we analyzed the presence of significant trends in burnt area, applying the contextual Mann–Kendall test and the Theil–Sen slope estimator. Data on burnt areas were obtained from the moderate-resolution imaging spectroradiometer (MODIS) burnt area product and the analyses were processed in TerrSet. Our results showed that ca. 30% of the country's area burned every year. The highest percentage of annual burnt area was found in northeast and southeast Angola, which showed large clusters of decreasing trends of burnt area. The clusters of increasing trends were found mainly in central Angola, associated with savannas and grasslands of Angolan Miombo woodlands. The protected areas of Cameia, Luengue-Luiana, and Mavinga exhibited large areas of decreasing trends of burnt area. Conversely, 23% of the Bicuar National Park was included in clusters of increasing trends. Distinct patterns of land cover were found in areas of significant trends, where the clusters of increasing trends showed a higher fraction of forest cover (80%) than the clusters of decreasing trends (55%). The documentation of burnt area trends was very important in tropical regions, since it helped define conservation priorities and management strategies, allowing more effective management of forests and fires in countries with few human and financial resources.

Keywords: wildfires; MODIS burnt area product; WWF ecoregions; land cover; Miombo woodlands; biodiversity conservation; sub-Saharan Africa

1. Introduction

Fire is considered a key driver of biodiversity [1], playing an important role in the structure and distribution of ecosystems and biochemical cycles [2]. It determines the distribution, ecology, and maintenance of African savannas and grasslands [3–5]. Over the last million years, the frequent occurrence of fires in Africa promoted the evolution of a more fire-tolerant and dependent flora [6]; without fire, large areas of savannas could develop into closed woodlands [4].

In southern Africa, however, the severity of fires caused by human activity is increasing [7], with negative consequences for biodiversity conservation and climate change. Approximately 70% of the global burnt area and 50% of fire-related carbon emissions are produced in Africa [8,9]. Uncontrolled fires can greatly damage residential zones, agricultural areas, and forests that are not adapted to fire, contributing to deforestation in many African countries, and producing gas emissions with harmful effects on human health and global climate [10–12]. In Africa, wildfires were identified as the main driver of forest degradation in remote areas of forests, which are more likely affected by recurring fires than by timber extraction [13]. On the other hand, fires are also used in the traditional slash-and-burn agriculture, which consists of cutting trees and woody plants from an area and then burning biomass, resulting in nutrient-rich ash that makes the soil more fertile for agriculture and grazing [13,14]. Savanna fires start by natural causes, mainly lightning, and human intervention, but the relative share of anthropogenic fires is quickly increasing in Africa. In savannas of West Africa, human activity is the most important factor causing fires [11].

The size of burnt areas varies greatly across countries and depends on rainfall seasonality, human activity, and available fuel [15,16]. The fire regime is strongly related to the land use practices and rural populations greatly influence the extent of fires and the annual area burnt [16,17].

Historically, humans have used fire to clean the land [16]. The rural population, which continues to grow in many African countries, frequently uses fire to promote grass growth for cattle to eliminate agricultural residues after harvests and to clear agricultural lands [16]. Moreover, wood and charcoal are the main sources of energy for cooking and heating in rural communities, their use enhancing the exploitation of natural resources and deforestation [11]. An increase in biomass burning causes local drying of the atmosphere, reducing rainfall in the region [12].

According to Archibald et al. [18], Angola, Zambia, and Mozambique are the countries with the most fire activity in southern Africa. In these countries, where the fires are mainly grass-fueled surface and closely related with areas of grass vegetation, such as savannas, grassland, forest and transitions, approximately 50% of the land area was affected by fire between 2001 and 2008, and a large part of it was burnt more than four times during this period [18].

Angola encompasses a great diversity of habitats from desert to tropical rain forests but is dominated by grasslands, savannas, and scrublands [19]. In total, 15 different ecoregions of the global ecoregions map of World Wildlife Fund (WWF) are present in Angola [20], of which the most widespread are the Miombo woodlands of the central plateau, and the western Congolian forest-savanna mosaics in the north [21]. Other important but less extensive ecoregions found in this country are the Atlantic Equatorial coastal forests in Cabinda and the mopane woodlands and the Namibian savanna woodlands in the southwest [20]. In this country, fires are generally classified as surface fires, occurring mainly during the dry season between late April and November [22,23]. Northern regions and central highlands burn considerably earlier than southern zones, when the grass fuel contains higher moisture, resulting in cooler, less intense fires that are less damaging to vegetation than the later and hotter fires in southern Angola [15,23].

Over the last years, intense fires are becoming more frequent in Angola [23]. Very dry seasons and consecutive years of drought with low-humidity conditions promote an increase in the occurrence and intensity of fire [24]. Moreover, droughts have led to significant food insecurity, reducing many populations to subsistence conditions and increasing use of fire to clear hunting zones and preparing new ones for cultivation [14]. Recent studies [23,25] show that Angolan forests have been strongly affected by severe and recurrent fires. Although a single fire has no strong effect on tree composition, the damage caused by recurrent fires in the late dry season can cause high tree mortality [26]. For woodlands in south Africa, a period of at least 5 years without fire is necessary for the successful reestablishment of some tree species [27,28], but such an interval rarely happens in Angolan savannas [17,29].

Currently, deforestation continues to increase in Angola, where one of the highest rates of tree cover loss in sub-Saharan Africa occurs [30–32]. Most of the native trees are under severe pressure due to logging, wood harvesting and charcoal production [21,33,34], as well as fire-related human and climate factors [35–37]. With southern Angola expected to become drier and warmer as a consequence of climate changes, fire frequency is likely to increase and the conservation of Angolan forests and savannas is urgently needed [38,39]. Effective biodiversity conservation requires that fire plays a natural role in ecosystems without representing a threat to ecosystems or human well-being [40].

The current forest degradation caused by fires is difficult to quantify, requiring a comparison of repeated measurements over time [26]. Over the last two decades, remote sensing has become a very useful technique to map burnt areas at regional and global scales, providing a comprehensive view of landscape patterns of fire (e.g., [10,18,29,41,42]). Mapping burnt areas over the years is crucial for resource managers to understand fire impacts on habitats, wildlife, and human settlements, particularly in the context of global climate change [11]. The moderate-resolution imaging spectroradiometer (MODIS) has been extensively used for the study of fire and burnt areas, including in southern Africa [18,24,25]. For instance, Archibald et al. [18] analyzed fire regimes in southern Africa, based on fire frequency, fire seasonality, fire radiative power, and fire size, as well as their relationship to human activities. Mishra et al. [24] examined how variations in burned area and fire frequency were determined by rainfall, vegetation morphology, and land use in the semiarid savanna of Botswana. Schneibel et al. [25] studied the role of fire in forest degradation processes in south-central Angola. However, the study of spatial and temporal trends of burnt area and the determination of its magnitude have never been carried out in Angola. MODIS derived burnt-area product (MCD64A1) is based on changes in surface reflectance over time and includes per-pixel burnt area and quality information about recent fires [43,44]. It is available as monthly global maps, with good detection of small burns, low temporal uncertainty, and a small extent of unmapped areas [18,44,45].

The main goal of this study is to analyze spatial and temporal trends of the area annually burnt in Angola between 2001 and 2019, and how these trends affected the conservation goals of the country, i.e., the sustainable use of natural resources, the recovery of its biodiversity, and the rehabilitation of its protected areas after almost 30 year of war [14]. Specifically, we aimed to: (i) map fire incidence in Angola between 2001 and 2019; (ii) detect and analyze positive and negative trends in the annual burnt area; and (iii) analyze the association between the detected trends and the Angolan ecoregions, protected areas, and land cover classes. Our results will provide new data and useful information for the sustainable fire management and conservation of Angolan ecosystems.

2. Materials and Methods

2.1. Study Area

With ca. 1,246,700 km^2, the Republic of Angola is the largest country in southern Africa. It is located between 4°22′ S and 18°02′ S, and 11°41′ E and 24°05′ E, and is organized into 18 provinces (Figure 1). The mean annual rainfall varies between ca. 50 mm in Namib and more than 1500 mm in Lunda Norte [46]. A very diverse climate and topography originated high species richness and habitat heterogeneity. In total, 15 ecoregions are recognized in Angola [20] (Figure S1), including deserts, savannas, grasslands, tropical forests, and mangroves. The current system of protected areas includes 14 areas (Figure 1), covering about 13% of the country [47,48].

With a population of over 26 million people and an annual growth rate of 3.3% according to the 2014 population census [49], this country struggles with food scarcity, poor medical support, and lack of infrastructures [25]. Approximately 37% of the population lives in rural areas and depends on agriculture for food and financial income [49,50]. The expansion of agricultural and urban areas and the demand for agricultural land are important causes of anthropogenic fire and deforestation [34,51].

Figure 1. Study area, with the 18 provinces of Angola and the protected areas system: BI, Bicuar; BU, Búfalo; KA, Cangandala; CM, Chimalavera; IO, Iona; IP, Ilheu dos Pássaros; KM, Cameia; KI, Quiçama; LL, Luengue-Luiana; LU, Luando; MV, Mavinga; MU, Mupa; MY, Maiombe; NA, Namibe.

2.2. Burnt Area Data, Trend Estimation, and Testing

Data on burnt areas in Angola were obtained from the MODIS burnt area product MCD64A1 v.006 [52] and downloaded using the Google Earth Engine platform. These data are produced from MODIS Terra and Aqua daily surface reflectance products at 500 m resolution [52–54]. Based on the daily surface reflectance, the algorithm identifies rapid changes to detect burning events and maps their spatial extent [52,54]. The algorithm uses a burn sensitive vegetation index derived from MODIS shortwave infrared atmospherically corrected surface reflectance to create thresholds that are applied to composite imagery data and then spatial and temporal active-fire information is used to estimate probabilistic thresholds suitable to classify individual grid cells as burnt or unburnt [44,52].

MODIS Terra and Aqua satellites were launched at different times. While the Terra sensor has been collecting data since February 2000, the Aqua sensor has been collecting data since June 2002 [55]. As the post-fire signal is persistent and the product of the burnt area is not very sensitive to the time of satellite passage [56], losing little information with just one sensor, we used monthly data from January 2001 to December 2019.

To analyze the mean annual fraction of the burnt area and the number of years of burning, we summed the data for the 12 months of each year (January to December) and resizes the resulting maps to a 5 km resolution grid using QGIS v.3.4.4 software [57]. Each cell was reclassified with the percentage of area burned and the number of years in which it was burned.

The analysis of the spatial and temporal trends of the burnt area was processed in TerrSet v.18.31 [58]. To avoid sequential correlation in time series (i.e., the lack of independence between observations), we applied the Durbin-Watson serial correlation test [59–61] and the pre-whitening procedure proposed by Wang and Swail [62], which maintains the same trend as the original series, but without serial correlation.

To identify significant trends, we applied the Mann–Kendall test [63,64] and the contextual Mann–Kendall test [65]. The Mann–Kendall test is a nonparametric test of trend monotonicity and a common approach in studies of environmental time series [61,65,66]. The contextual Mann–Kendall test, proposed by Neeti and Eastman [65], is a modified version of the Mann–Kendall test that includes geographical contextual information, i.e., neighboring cells are involved in the determination of trends, allowing the detection of homogeneous regions with similar trends. The contextual analysis imparts greater confidence to the identified trends [61].

Then, the Theil–Sen slope estimator was applied to determine the rate of burnt area change in each cell with significant trends [67,68]. This non-parametric technique is robust against outliers and provides an indicator of change over time [66]. It is defined as the median of the slopes calculated between observations in all pair-wise time steps and is suitable for application to time series with large annual variations, such as those concerning annual burnt area [61]. An increasing trend is indicated by a positive Z value, while a decreasing one is indicated by a negative Z value.

2.3. WWF Ecoregions, Protected Areas, and Land Cover Data

Ecological conditions, changes in land use, and population dynamics are important drivers of fire occurrence over time. To analyze the association between these factors and the trends of the burnt areas, we compared the maps of burnt area trends with the WWF ecoregions (Figure 2a) [20], the Angolan system of protected areas [47], and land cover time-series maps (Figure 2b) [69,70].

WWF terrestrial ecoregions represent the high variability of biogeographic diversity and ecological conditions (see Figure 2a) [20,71–73]. This map was produced at a global scale and reflects the regionalization of the Earth's terrestrial biodiversity, defining units that contain a distinct assemblage of natural communities sharing a large majority of species, dynamics, and environmental conditions [20]. WWF ecoregions has been used as a basis for studies on biotic diversity and conservation in Angola [19,21,74].

Maps concerning the Angolan system of protected areas were downloaded from the World Database of Protected Areas [75]. The present extent of Luengue-Luiana and Mavinga national parks was not updated in WDPA, so it was vectorized based on published Angolan legislation (Diário da República de Angola, law 38/11 29 December 2011, p. 6340).

Land cover data were obtained from the European Space Agency (ESA) Climate Change Initiative Land Cover (CCI-LC) project [76] which produced annual land cover maps for 1992–2015 (version 2.0.7) and 2016–2018 (version 2.1.1), combining remote sensing products and ground observations [69,70]. These maps provide data describing the land surface at 300 m resolution, divided into 22 classes (see Figure 2b). To understand changes in land cover over time, we used the most recent land cover data available (2018) and data collected every four years (2002, 2006, 2010, and 2014). Finally, the maps of burnt area and significant trends were spatially intersected with the WWF ecoregions, protected areas, and land cover maps using QGIS v.3.4.4 software [57].

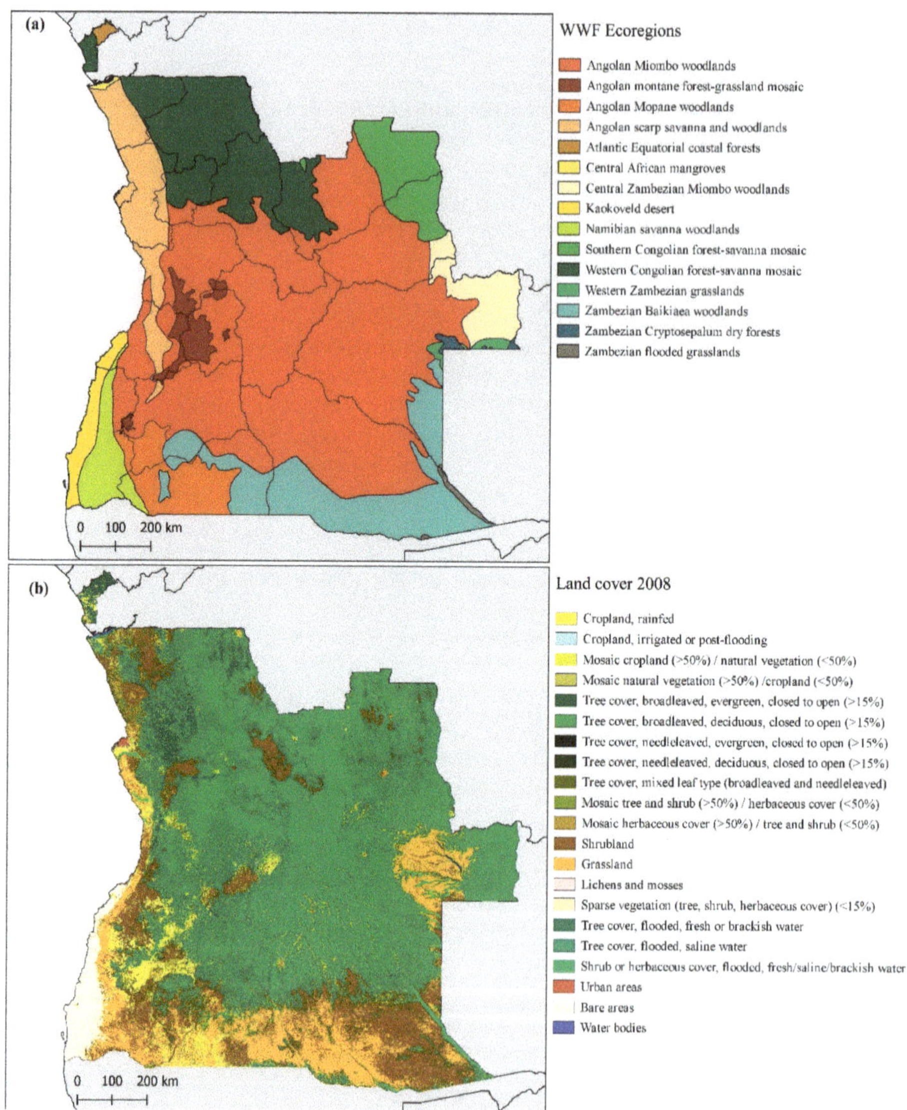

Figure 2. World Wildlife Fund (WWF) ecoregions and land cover map of Angola: (**a**) WWF ecoregions occurring in Angola [20]. The black lines represent the boundaries of the 18 provinces of Angola; (**b**) land cover map of Angola in 2018 (300 m resolution) adapted the European Space Agency (ESA) Climate Change Initiative Land Cover (CCI-LC) project [70].

3. Results

3.1. Area Burnt in Angola since 2001

Our results revealed that Angola has been extensively affected by fires every year. On average, ca. 368,300 km^2 burn every year since 2001, corresponding to 30% of the country's area. The highest values of burnt area were recorded between 2003 and 2005, with more than 400,000 km^2 burnt each year, and the lowest value was 315,580 km^2 in 2018 (Figure 3a).

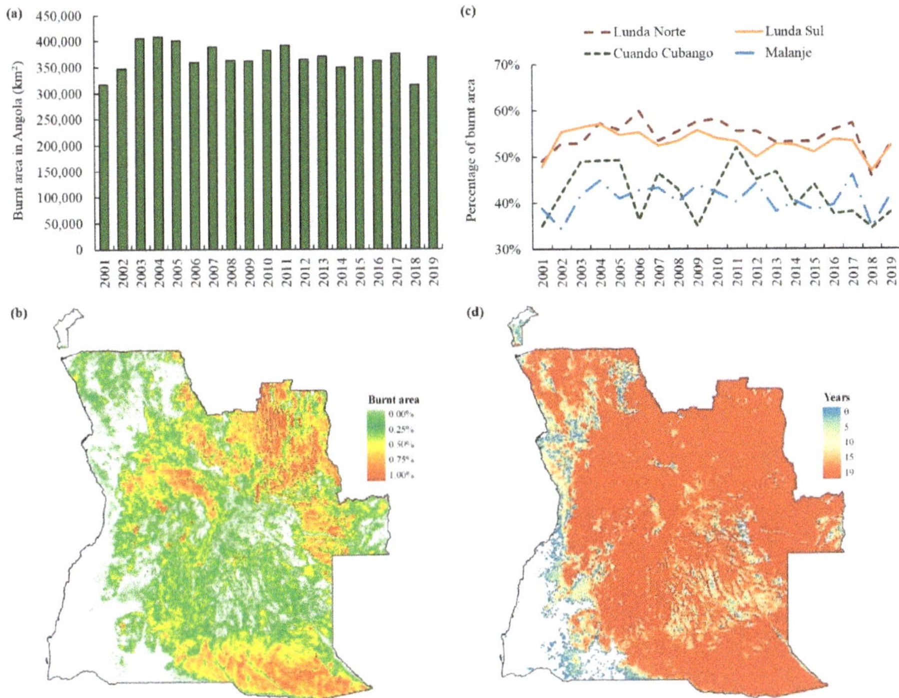

Figure 3. Burnt area in Angola, from 2001 to 2019: (**a**) Absolute burnt area (km^2) per year; (**b**) map of the mean annual fraction of burnt area in each cell (5 km resolution); (**c**) percentage of the area burnt in the four provinces with the highest values; (**d**) map of the number of years when each cell was completely or partially burnt since 2001 (5 km resolution).

The northeast and southeast regions of Angola presented the highest percentage of annual burnt area per cell (Figure 3b). The provinces with the largest annual burnt areas were Lunda Norte (54%, ca. 58,840 km^2), Lunda Sul (53%, ca. 41,375 km^2), Cuando Cubango (42%, ca. 84,455 km^2), and Malanje (41%, ca. 33,586 km^2) (Figure 3c, Table S1). Among them, Cuando Cubango showed the highest variations over consecutive years. The most affected municipalities were Lubalo, Cuilo, and Lucapa in Lunda Norte; Saurimo and Muconda in Lunda Sul; Dirico and Rivungo in Cuando Cubango; and Luquenbo and Cangandala in Malanje. Intermediate values were found in Cuanza Sul, Moxico, Bié, Cunene, Cuanza Norte, Huambo, Zaire, Huíla, Uíge, and Benguela, with 12–30% of area burnt every year. The provinces of Bengo, Cabinda, Luanda, and Namibe had the lowest values, under 10%. Fire events are recurrent in Angola, except in the arid southwestern region. Figure 3d shows the number of years when each cell burnt totally or partially between 2001 and 2019. Since 2001, 64% of the cells burnt 15 years or more.

3.2. Spatial Trends of Burnt Area in Angola since 2001

The results of the contextual Mann–Kendall test highlights large areas of significant increasing and decreasing trends in the burnt area of Angola. Figure 4 shows the results of this test, where the red cells represent increasing trends (positive Z value) and green cells represent decreasing trends (negative Z values). The significant areas with a confidence level of 95% are outlined with a black line. The Mann–Kendall test was also performed (Figure S1), but the contextual Mann–Kendall test (Figure 4) performed better in creating larger and more homogenous clusters of cells corresponding to significant areas.

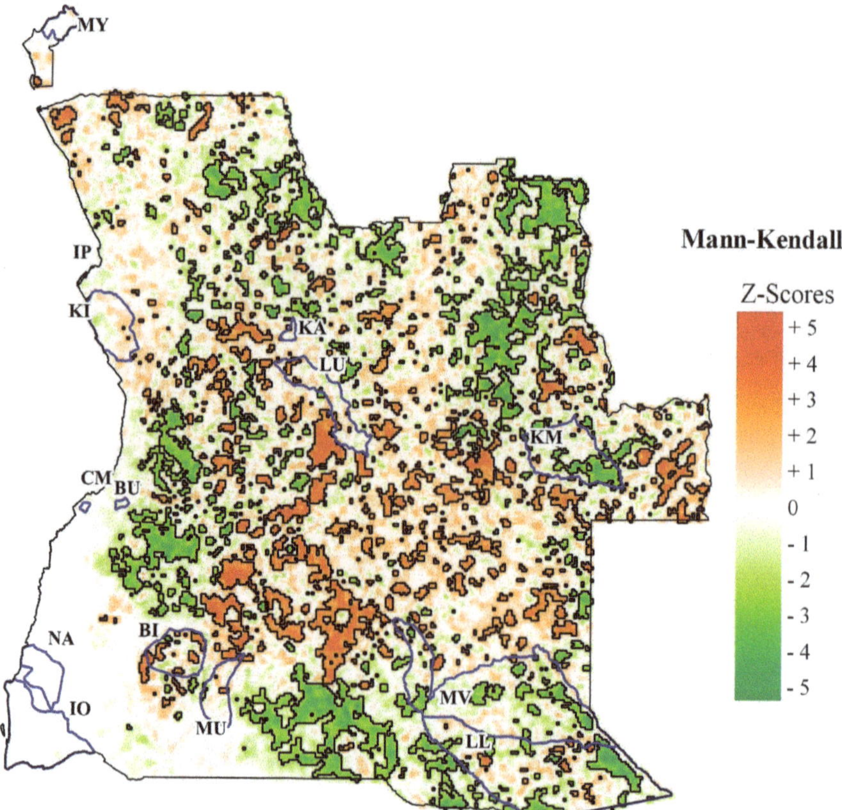

Figure 4. Trends of the annual burnt area in Angola for the 2001–2019 period, based on the contextual Mann–Kendall test. The positive z-scores correspond to areas of increasing trends of burnt area, while the negative z-scores correspond to decreasing trends. The significant areas (with a confidence level of 95%) were outlined with black lines. Protected areas are indicated with blue outlines: BI, Bicuar; BU, Búfalo; KA, Cangandala; CM, Chimalavera; IO, Iona; IP, Ilheu dos Pássaros; KM, Cameia; KI, Quiçama; LL, Luengue-Luiana; LU, Luando; MV, Mavinga; MU, Mupa; MY, Maiombe; NA, Namibe.

We detected significant areas of increasing trends in ca. 119,410 km^2 (10% of the country's area), mainly in the central region of the country. With more than 9400 km^2, the single largest area is located in the north of the Cuando Cubango province, and the second largest area, with ca. 4330 km^2, is found in Bié. The provinces with the most extensive areas of significant positive trends are Moxico (34,731 km^2), Cuando Cubango (18,628 km^2), Bié (15,830 km^2), and Huíla (12,658 km^2) (Figure 5, Table 1). Significant areas of decreasing trends was detected in ca. 154,240 km^2 (12% of the country's area), spread all across Angola, except in its central region. The single largest area has more than 21,800 km^2 and covers part of the Cuando Cubango and Cunene provinces; the second largest single area has approximately 14,000 km^2 and is located in Lunda Sul. The provinces with the most extensive areas of significant negative trends are Cuando Cubango (34,584 km^2), Lunda Sul (21,049 km^2), and Lunda Norte (19,353 km^2) (Figure 5, Table 1).

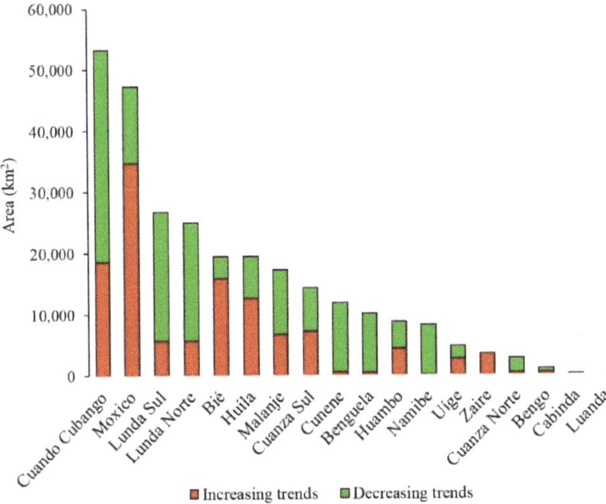

Figure 5. Areas of significant increasing and decreasing trends in Angolan provinces.

Table 1. Area and proportion of increasing and decreasing significant trends of annual burnt area in Angolan provinces, WWF ecoregions, and protected areas, from 2001 to 2019.

Provinces	Total Area	Area of Decreasing Trend		Area of Increasing Trend	
	(km²)	(km²)	(%)	(km²)	(%)
Bengo	34,363	579	1.7	425	1.2
Benguela	39,509	9689	24.5	459	1.2
Bié	72,048	3625	5.0	15,830	22.0
Cabinda	7119	0	0.0	188	2.6
Cuando Cubango	199,483	34,581	17.3	18,628	9.3
Cuanza Norte	23,823	2301	9.7	381	1.6
Cuanza Sul	55,257	7196	13.0	7149	12.9
Cunene	77,259	11,347	14.7	545	0.7
Huambo	33,133	4391	13.3	4303	13.0
Huíla	78,684	6769	8.6	12,658	16.1
Luanda	2447	0	0.0	0	0.0
Lunda Norte	107,973	19,353	17.9	5639	5.2
Lunda Sul	77,927	21,049	27.0	5680	7.3
Malanje	82,163	10,645	13.0	6641	8.1
Moxico	199,986	12,506	6.3	34,731	17.4
Namibe	57,911	8100	14.0	82	0.1
Uíge	62,005	2102	3.4	2610	4.2
Zaire	36,590		0.0	3461	9.5
WWF ecoregions					
Angolan Miombo woodlands	628,703	64,488	10.3	86,310	13.7
Angolan montane forest-grassland mosaic	25,419	8251	32.5	1172	4.6
Angolan mopane woodlands	51,064	1344	2.6	1045	2.0
Angolan scarp savanna and woodlands	73,947	3375	4.6	2552	3.5
Atlantic Equatorial coastal forests	2534	0	0.0	0	0.0
Central African mangroves	1240	0	0.0	326	26.3
Central Zambezian Miombo woodlands	40,648	1769	4.4	6633	16.3
Kaokoveld desert	20,590	0	0.0	0	0.0
Namibian savanna woodlands	33,491	0	0.0	0	0.0
Southern Congolian forest-savanna mosaic	58,558	18,693	31.9	3208	5.5
Western Congolian forest-savanna mosaic	168,864	23,118	13.7	9698	5.7
Western Zambezian grasslands	4606	133	2.9	1134	24.6
Zambezian Baikiaea woodlands	131,010	32,382	24.7	6559	5.0
Zambezian Cryptosepalum dry forests	3085	0	0.0	776	25.1
Zambezian flooded grasslands	3197	681	21.3	0	0.0

Table 1. *Cont.*

Provinces	Total Area	Area of Decreasing Trend		Area of Increasing Trend	
	(km^2)	(km^2)	(%)	(km^2)	(%)
Protected Areas					
Bicuar National Park	7728	162	2.1	1809	23.4
Bufalo Partial Reserve	332	0	0.0	0	0.0
Chimalavera Natural Regional Park	214	0	0.0	0	0.0
Ilha dos Pássaros Integral Nature Reserve	0	0	0.0	0	0.0
Iona National Park	15,264	0	0.0	0	0.0
Cameia National Park	14,185	3918	27.6	161	1.1
Cangandala National Park	642	0	0.0	33	5.1
Quiçama National Park	8597	0	0.0	42	0.5
Luando Integral Nature Reserve	8737	48	0.5	769	8.8

3.3. Burnt Area Trends According to WWF Ecoregions, Protected Areas, and Land Cover

Our results revealed a very uneven distribution of burnt area trends across WWF ecoregions (Figure 6a, Figure S2, and Table 1). Central African mangroves and Zambezian Cryptosepalum dry forests are among the smallest ecoregions occurring in Angola, but they displayed the highest percentage of increasing trends, and no significant decreasing trends. Western Zambezian grasslands, Central Zambezian Miombo woodlands, and Angolan Miombo woodlands also showed very high percentages of increasing trends (25%, 16%, and 14%, respectively) and much lower percentages of decreasing trends (3%, 4%, and 10%, respectively). The Angolan Miombo woodlands is the largest ecoregion (more than 628,000 km^2), covering the central region of Angola.

The high percentage of decreasing trends was found in the Angolan Montane forest-grassland mosaic (32%), the Southern Congolian forest-savanna mosaic (32%), the Zambezian Baikiaea woodlands (25%), the Zambezian flooded grasslands (21%), and the Western Congolian forest-savanna mosaic (14%). The Namibian Savanna woodlands, Kaokoveld desert, and Atlantic Equatorial coastal forests exhibited no significant trends of burnt area.

In protected areas, burning trends are mainly negative. Our results revealed areas of decreasing trends in five national parks, totaling ca. 19,560 km^2, which represents 12% of the national protected area (Figure 4, Table 1). Cameia has the highest percentage of decreasing trends (28%, 3918 km^2), followed by Luengue-Luiana (18%, 7811 km^2) and Mavinga (17%, 7466 km^2) national parks (Figure 6b). The area of increasing trends was ca. 5710 km^2, corresponding to 4% of the global protected area. It was detected in eight protected areas and the highest proportion was found in Bicuar National Park (23% of the park's area, 1809 km^2) (Figure 6b, Table 1). Luando reserve (9%, 769 km^2), Cangandala (5%, 33 km^2), Mupa (5%, 240 km^2), and Luengue-Luiana (4%, 1949 km^2) national parks also have important fractions of increasing trends. Some national parks, such as Namibe, Iona, and Maiombe, showed no significant trend.

Concerning the land cover, Angola is covered mainly by natural vegetation as forest, shrubland, and grassland. In 2018, forest (including broadleaved, deciduous, evergreen, and flooded tree classes) was found in 59% of the country's area, shrubland in 18% and grassland in 12% (Figure 7a). Cropland was identified in ca. 48,900 km^2, representing only 4% of the territory. Comparing the composition of the areas included in the significant clusters, clusters with increasing trends presented a much higher fraction of forest classes (80%) than clusters with decreasing trends (55%). In clusters with decreasing trends, the proportions of grassland and shrubland were much higher (20% and 19%, respectively) than in the increasing ones (4% and 10%, respectively) (Figure 7a and Figure S3).

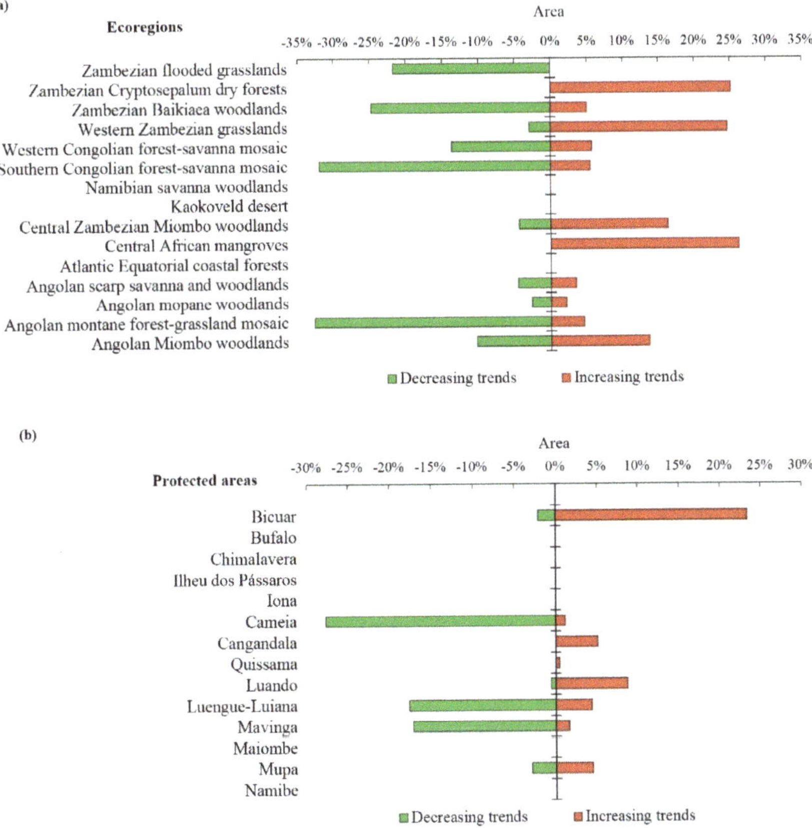

Figure 6. Areas (%) of significant increasing and decreasing trends: (**a**) in WWF ecoregions; (**b**) in protected areas.

Areas with significant increasing and decreasing trends show similar patterns of land cover changes since 2002 (Table 2), occurring a global decrease in natural vegetation extent, i.e., grassland, shrubland, and forest, as well as an increase in cropland and urban areas (Figure 7). In significant clusters, forest shows a small increase in area between 2002 and 2006, but after 2006 it decreases and reaches the lowest value in 2018. Shrubland has maintained almost stable since 2010, with a small increase during the last four years in clusters with decreasing trends. Between 2002 and 2018, clusters with increasing trends have lost of 0.04% of forest, 0.38% of shrubland, and 1.26% of grassland, but displayed an increase of cropland (8.84%). Decreasing trend clusters lost 0.13% of forest area, 0.40% of shrubland and 1.19% of grassland, and has a great increase of cropland by 15.87% (Figure 8). Urban areas represented a small fraction inside the clusters, but they have a great increase, mainly in clusters with decreasing trends.

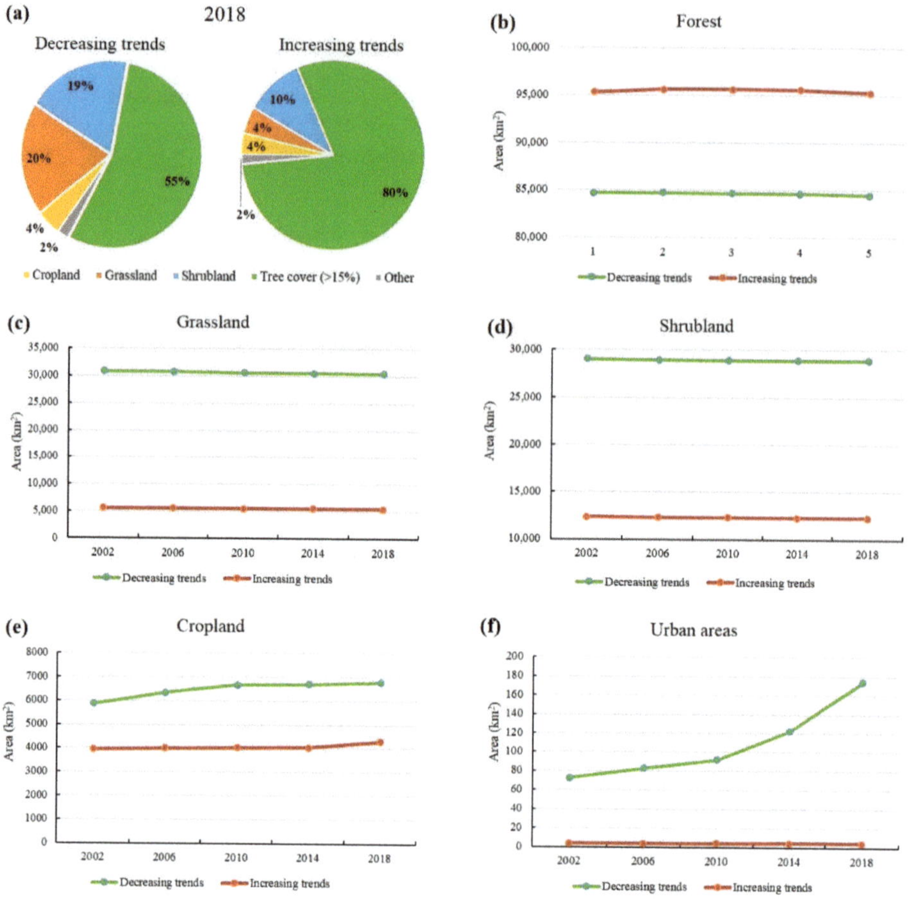

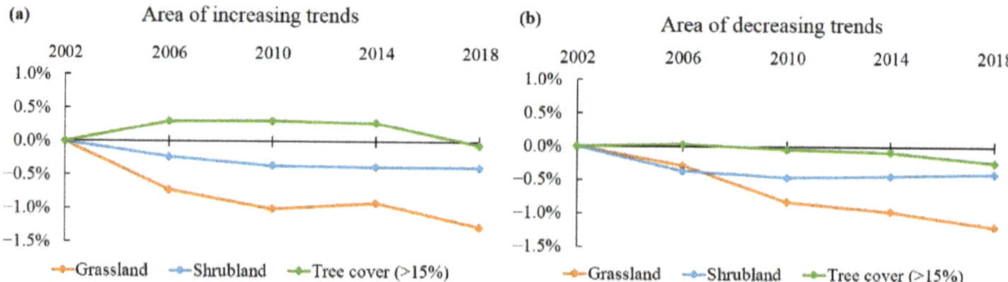

Figure 7. Land cover and the significant trends of burnt areas from 2002 to 2018, in increasing and decreasing trends clusters: (**a**) Proportions of the main land cover classes in 2018 (the last year for which information is available); (**b**) area of forest; (**c**) area of grassland; (**d**) area of shrubland; (**e**) area of cropland; (**f**) area of urban areas.

Figure 8. Changes in area of natural land cover classes (grassland, shrubland and forest) since 2002 to 2018, (**a**) in increasing trends clusters, and (**b**) in decreasing trend clusters.

Table 2. Area of land cover classes within the significant clusters of burnt area trends, in 2002, 2006, 2010, 2014, and 2018.

Land Cover Class	Area of Decreasing Trends						Area of Increasing Trends					
	2002 (km²)	2006 (km²)	2010 (km²)	2014 (km²)	2018 (km²)	Change 2001–2018 (km²)	2002 (km²)	2006 (km²)	2010 (km²)	2014 (km²)	2018 (km²)	Change 2001–2018 (km²)
Bare areas	0	0	0	0	0	0	2	2	2	2	2	0
Cropland, irrigated, or post-flooding	2	2	2	2	3	2	2	2	2	2	3	1
Cropland, rainfed	4153	4425	4581	4606	4656	503	2203	2239	2251	2251	2349	146
Grassland	30,162	30,081	29,904	29,859	29,795	−366	4938	4939	4928	4932	4942	4
Mosaic cropland (>50%)/natural vegetation (tree, shrub, herbaceous cover) (<50%)	1720	1925	2094	2111	2148	428	1746	1760	1777	1792	1948	202
Mosaic herbaceous cover (>50%)/tree and shrub (<50%)	706	701	710	711	705	−1	621	580	575	576	547	−74
Mosaic natural vegetation (tree, shrub, herbaceous cover) (>50%)/cropland (<50%)	1759	1483	1382	1379	1394	−365	1039	693	673	659	684	−355
Mosaic tree and shrub (>50%)/herbaceous cover (<50%)	1054	1008	1015	1015	1032	−22	603	620	632	653	682	78
Shrub or herbaceous cover, flooded, fresh/saline/brackish water	467	471	484	481	497	30	441	440	446	447	458	17
Shrubland	29,043	28,936	28,910	28,916	28,927	−116	12,364	12,336	12,320	12,318	12,317	−47
Sparse vegetation (tree, shrub, herbaceous cover) (<15%)	83	83	83	83	83	0	0	0	0	0	0	0
Tree cover, broadleaved, deciduous, closed to open (>15%)	82,910	82,938	82,870	82,837	82,696	−214	94,565	94,827	94,829	94,801	94,493	−72
Tree cover, broadleaved, evergreen, closed to open (>15%)	1753	1753	1756	1758	1770	17	762	789	788	790	784	22
Tree cover, flooded, fresh, or brackish water	51	52	53	52	52	1	8	12	13	13	13	5
Tree cover, flooded, saline water	0	0	0	0	0	0	10	13	13	13	14	4
Urban areas	73	83	92	123	174	101	4	4	4	5	5	1

4. Discussion

4.1. Trends of Annual Burnt Area in Angolan Provinces and Ecoregions

This study presents a spatial and temporal analysis of the area burnt in Angola based on MODIS satellite images collected from 2001 to 2019, which allows us to assess the main trends over the years. Our results unveiled new information about fire in Angola, revealing the main areas of increasing and decreasing trends of the annual burnt area. There are several studies on fire regimes and dynamics at the global scale or focused in southern Africa (e.g., [17,18,42,77,78]); however, this is the first study on fire trends carried out specifically in Angola and focused on its ecosystems and protected areas. This country encompasses unique habitats and species, and is possibly one of the least documented biodiversity hotspot areas in the world [79]. However, with continued population growth and lack of financial support, subsistence agriculture and human pressure on the land will continue to increase, putting high pressure on natural resources [19,34,51] and endangering the conservation of the Angolan natural heritage [37,47].

The highest values of burnt area were recorded between 2003 and 2005, which may be a consequence of the end of the civil war. During the civil war, ca. one million people were killed and over four million people displaced. Moreover, there was a massive migration from rural communities to the cities and fields were abandoned [80,81]. When the war ended, some displaced people returned to rural areas and burnt natural vegetation to clear land for agriculture.

The largest burnt areas and the highest frequency of fire were found in grasslands and savannas of Lunda Norte, Lunda Sul, and Malanje (northeast Angola), as well as in open savannas of Cuando Cubango (southeast Angola). This agrees with previous studies that recognized these regions as the most affected by fires in Angola [15,17,23,77]. However, these provinces presented small areas of increasing trends and the largest clusters of decreasing trends of burnt area, suggesting a possible decrease in fire extent and/or frequency in the coming years, if the current trends are maintained. Lunda Norte and Lunda Sul are formed mainly by vast plains drained by the tributaries of the Cassai–Congo Basin [46]. According to the Köppen–Geiger climatic classification, the climate is humid tropical and tropical savanna, with hot and humid summers, and warm and dry winters [82]. The population density and the practice of agriculture are very low, but the wind plays an important role in the occurrence of fires, drying the grasslands [49,83]. Cuando Cubango is composed mainly of Kalahari sands, as it is a vast plain with slow flowing rivers. The climate is humid subtropical in the north and hot semi-arid in the south, with a low average annual rainfall, ranging from 1000 to 600 mm [46,82]. The population density is low in this region but more than half of the people work in agricultural activities [49]. Malanje has a similar pattern, with low population density but high intensity of agricultural activities [4]. This province includes a plateau in its central region, with elevation from 1000 to 1250 m, and the climate is tropical savanna [46,82].

The main areas of increasing trends were found in central Angola. This region includes Cuanza Sul, Bié, Huambo, Moxico, and Huíla provinces, and consists mainly of Angolan Miombo woodlands. Moxico is located in central-eastern Angola, the mean annual rainfall ranges from 900 to 1300 mm and is dominated by a humid subtropical climate. Cuanza Sul, Bié, Huambo, and Huíla are located in the extensive region of the ancient plateau, characterized by high levels of annual precipitation in the central zone, reaching 1300 mm in the province of Huambo. [46,49,82]. These provinces are dominated by tropical savanna, humid subtropical, and oceanic climates [82]. Currently, they have a very high population density and a high fraction of families working in agriculture. The Angolan Miombo woodlands is an ecoregion almost exclusive to Angola, differing from other savannas and forests in its high abundance of trees of the Leguminosae family, specifically, *Brachystegia*, *Julbernardia*, and *Isoberlinia* [72,84]. Several authors recognized fire as one of the most important ecological factors in Miombo, preserving the balance between grasses and trees [11,85]. Most plant species of Miombo show some degree of fire resistance, yet they cannot survive intense and recurrent fires [11]. The natural cause of Miombo fires is mainly thunderstorms in the early rainy season, but most fires are caused by human action and occur during the dry and hot season [11], when the grass fuel is drier, resulting in more intense, damaging fires. A continuous increase in burnt area for a long period could represent a serious threat for tree species, transforming Miombo woodlands into open savannas [23].

The Central Zambezian Miombo woodlands should also be a priority for future studies and fire management measures. The ecoregion is found in the eastern province of Moxico and show more than 16% of the area with significant increasing trends. Both ecoregions, the Angolan Miombo woodlands and the Central Zambezian Miombo woodlands, are included in the regional center of Zambezian endemism (defined by White [86]), housing a high richness and endemism of plant and animal species [72]. Fire is an integral part of the Miombo ecology, but changes in fire regimes, such as the increase in frequency or severity, can result in drastic landscape changes, reducing or removing some species. Zambezian Cryptosepalum dry forests, Western Zambezian grasslands, and Central African mangroves showed the highest percentage of increasing trends. However, these ecoregions are very little represented in the country, with less than 5000 km^2 each, and these results cannot be considered representative. Mangroves are present in a small area on the bank of the Congo River, surrounded by extensive areas of Angolan scrap savanna and woodlands. The cluster of growing trends detected in the mangrove area is shared with an extensive area of scrap savanna, suggesting that it may correspond to a commission error, resulting from the effect of the neighboring cells involved in the determination of trends. The same may be occurring with Zambezian Cryptosepalum dry forests and Western Zambezian grasslands, these ecoregions are surrounding by Angolan and Central Zambezian Miombo woodlands, which has large clusters of significant increasing trends of burnt area in transitional zones.

The WWF ecoregions show very clear patterns of increasing and decreasing trends in the area burnt. The ecoregions represent the original distribution of distinct assemblages of species and communities [20], better distinguishing the structural organization of the landscape than the generic vegetation classes. The definition of fire management measures and laws must be adapted to each ecoregion's characteristics, addressing socioeconomic and ecological problems that restore, protect, and maintain natural habitats.

4.2. Implications for Vegetation and Land Cover

The largest clusters of decreasing trends in burnt area are found in the regions most affected by fire events every year, such as Lunda Norte, Lunda Sul, and Cuando-Cubango. These regions have a lower percentage of forest (ca. 55%) and a large percentage of shrubland (19%) and grassland (20%), a pattern probably resulting from the recurrent fires identified in these provinces that limit the growth of trees and shrubs. In these clusters, there was also a strong increase in the extent of urban areas, which more than doubled from 2002 to 2018, and of cropland areas, which slowly but continuously has been growing over the years.

On the other hand, the areas of increasing trends are characterized by a large proportion of forest (ca. 80%) and small fractions of shrubland (10%) and grassland (4%). These areas also have a relatively high proportion of cropland (4%), which substantially increased since 2004. In these regions, the continuous increasing trend of the annual burnt area may lead to the conversion of forests and woodlands into grasslands and shrublands. Interestingly, Cuando Cubango is among the provinces that displays more areas of significant positive trends but also more areas of negative trends. This reveals high turnover of burnt areas in the province, possibly resulting in high levels of land cover changes.

Our findings show a reduction in forest cover over the years, which is in agreement with other studies [23,42], confirming that land cover changes in Africa have been dominated by woodland and forest losses. The fires have had severe effects on Angolan woodlands, converting large areas of forest and woodlands into open land [25]. For instance, a study by Palacios et al [87] reported that 78.4% of Huambo province was covered by Miombo woodland in 2002, but 13 years later it had dropped to 48.3%. Similar losses in western Cuando Cubango, eastern Huíla, and eastern Huambo were reported by Schneibel et al. [88], and Mendelsohn and Mendelsohn [89]. However, the rate of forest loss detected in this study appears to be much lower than the rate reported by Hansen et al. [30], as reported by Food and Agriculture Organization of the United Nations (FAO) [31,90] and Global Forest Watch [32]. Other studies of land cover change [91,92] have also detected a much lower estimate of total forest area and forest loss when using ESA CCI-LC maps. One possible explanation is the different data sources. FAO collected forest resource data from the member countries [31], using satellite imagery and field surveys in different countries, while Hansen et al. [30] used high spatial resolution satellite data (30 m) obtained from NASA's Landsat, which allows for the detection of small-scale forest clear cuts. The different fraction of canopy considered in the forest definitions is also a possible cause of the disparity in the results. FAO [31] adopted a definition >10% canopy cover, while Global Forest Watch defined >30% canopy cover as forest. In our study, forest is defined as areas with >15% canopy cover, which does not reflect the decrease in the percentage of tree cover within each area, unless it falls below 15% of coverage.

The expansion of shifting cultivation practiced by smallholders and the overexploitation for timber, fuelwood, and charcoal are possibly the most important factors driving the rapid loss of Angolan woodlands and forests, as well as their very slow recovery rate [23]. The relative lack of nutrients and moisture in soils forces the farmers to leave their fields and use fire to clear new fields and enrich the soil with nutrients from the ash of the woody biomass [93]. The rate at which the woodlands are cleared and converted into cropland is dependent on the need to feed a growing human population. To reduce the impacts of human activity on fire processes, management strategies should be urgently implemented. These strategies must incorporate the ecological and socio-economic roles of fire, as well as the conservation of Angolan ecosystems [16].

4.3. Conservation and Management of Protected Areas

The protected areas system shows a strong impact on the burnt areas, presenting larger clusters of decreasing trends than of increasing trends. For instance, Cameia, Luengue-Luiana, and Mavinga national parks are located in eastern Angola (Moxico and Cuando Cubango provinces) and have the largest areas of decreasing annual burnt area. The decreasing trends detected in these areas may be related to the increase in the number of herbivores after the civil war. For instance, a study carried out in Luiana National Park showed a rapid increase in the numbers of African elephants between 2004 and 2005 [94]. According to Goldammer and Ronde [11], the abundance of mammalian herbivores is an important determinant of fire intensity. When the number of herbivores is high, less fuel is available and fires can affect smaller areas, with lower intensities [11,95].

Bicuar National Park is the most worrying protected areas, with increasing trends in more than 23% of its area. It is the only protected area of Huíla, located in a transition zone between the Angolan Miombo woodlands and the Zambezian Baikiaea woodlands ecoregions. With an area of almost 8000 km^2, it consists mainly of riverbanks of the Cunene River, sandy hills, and river valleys with savannah grasslands. It was created in 1938 as a hunting reserve due to the abundance of game species, such as African bush elephant (*Loxodonta africana*), blue wildebeest (*Connochaetes taurinus* subsp. *taurinus*), and eland (*Tragelaphus oryx*) [96]. The analysis of Luando Integral Nature Reserve also reveals significant increasing trends in ca. 9% of this area. The status of protected area was acquired in 1955, to conserve the giant sable antelope (*Hippotragus niger* subsp. *variani*) [97], including more than 8000 km^2 of the Angolan Miombo woodlands.

Angolan protected areas need a real and efficient administration, in addition to integrated biodiversity management plans. In 2006, the National Parks of Quissama, Mupa, Cangandala, Iona, Cameia, and Bicuar were almost completely abandoned, without equipment or supervision [98]. Parts of these protected areas are occupied by human populations that hunt and burn without control, leading to the disappearance of large mammals and converting large areas into agricultural land [78]. However, the Protected Areas system was greatly expanded in 2011 and more resources are being made available for more effective management [46]. The difficulties imposed by limited budgets, scarce human resources, and weak technical capacity impose the need to identify priority areas, where the government's conservation resources must be allocated first.

Fires have been the main management tool used in protected areas of the African savanna [95]. During the 20th century, managers of protected areas have considered both the suppression of unwanted fires and ignition of prescribed fires to achieve different goals, such as reducing the putative negative effects of fire on ecosystems, reducing the risk of unplanned fires, improving visibility for game viewing and policing, improving forage quality, and fighting invasive alien species [95,99–101]. Fire managers must decide how, where, and when to apply fires to achieve their goals of management, while also respecting legal constraints. During the 21st century, fire management in Angola will face new challenges imposed by the effects of climate change and a growing human population.

5. Conclusions

In this study, we detected spatial and temporal patterns of burnt areas across Angola, which is a crucial step for the definition of conservation priorities and management strategies. The identification of the areas with significant trends is extremely important in tropical areas with low field data available and few human and financial resources, as is the case of Angola. This will allow for efficient distribution of fire management resources and public investment, resulting in improved management of Angolan vegetation.

Understanding the role of climatic and anthropogenic factors on these trends is very important. The overexploitation of forests for the production of timber, firewood, and charcoal also enhances grass development, which in turn promotes more intense fires in the dry season. Effective actions to limit fire events are urgently needed to protect fire-sensitive resources, natural ecosystems, and to reduce carbon emissions to the atmosphere. While most southern African countries have legislation and regulations to use and control fire, these are rarely applied due to enforcement difficulties [11]. Fire control is also difficult because, in Africa, fires occur in thousands of small, dispersed events. Moreover, climate change will bring more frequent and intense droughts to Angola, with strong impacts on water resources, agricultural productivity, and wildfire potential. All of these factors are expected to play an important-negative-role on the trends of burnt areas, land cover change, and biodiversity conservation.

Supplementary Materials: The following are available online at http://www.mdpi.com/1424-2818/12/8/307/s1, Figure S1: Trends of annual burnt area in Angola for the 2001–2019 period, based on Mann–Kendall single-cell analysis. The positive z-scores correspond to areas of increasing trends of burnt area, while the negative z-scores correspond to decreasing trends. The significant areas (with a confidence level of 95%) were overlaid with black lines. Protected areas are limited with blue lines: BI, Bicuar; BU, Búfalo; KA, Cangandala; CM, Chimalavera; IO, Iona; IP, Ilheu dos pássaros; KM, Cameia; KI, Quiçama; LL, Luengue-Luiana; LU, Luando; MV, Mavinga; MU, Mupa; MY, Maiombe; NA, Namibe. Figure S2: WWF ecoregions occurring in Angola and the significant cluster of increasing and decreasing trends of annual burnt area detected in 2001– 2019 period, based on the Contextual Mann–Kendall test. Figure S3: Land cover map of Angola in 2018 (300 m resolution), adapted the European Space Agency (ESA) Climate Change Initiative Land Cover (CCI-LC), and the significant cluster of increasing and decreasing trends of annual burnt area detected in 2001–2019 period, based on the contextual Mann–Kendall test. Table S1: Estimated fraction of burnt area per year and the average of the burnt area from 2001 to 2019 in each province.

Author Contributions: Conceptualization, S.C., M.M.R. and J.M.C.P.; methodology, S.C., J.M.N.S. and J.M.C.P.; formal analysis, S.C., V.A., J.M.N.S. and J.M.C.P.; writing—original draft preparation, S.C.; writing—review and editing, M.M.R., R.F., V.A., J.M.N.S. and J.M.C.P. All authors have read and agreed this version of the manuscript.

Funding: This research was funded by Foundation for Science and Technology (FCT) of the Portuguese Government through the grants SFRH/BD/120054/2016 to S.C., UID/AGR/04129/2019 to Linking Landscape, Environment, Agriculture and Food (LEAF), UID/BIA/00329/2019 to Centre for Ecology, Evolution and Environmental Changes (cE3c), UIDB/00239/2020 to Forest Research Centre (CEF).

Conflicts of Interest: The authors declare no conflict of interest. The funders had no role in the design of the study; in the collection, analyses, or interpretation of data; in the writing of the manuscript, or in the decision to publish the results.

References

1. He, T.; Lamont, B.B.; Pausas, J.G. Fire as a key driver of earth's biodiversity. *Biol. Rev.* **2019**, *94*, 1983–2010. [CrossRef]
2. Kahiu, M.N.; Hanan, N.P. Fire in sub-Saharan Africa: The fuel, cure and connectivity hypothesis. *Glob. Ecol. Biogeogr.* **2008**, *27*, 946–957. [CrossRef]
3. Higgins, S.I.; Bond, W.J.; Trollope, W.S.W. Fire, resprouting and variability: A recipe for grass-tree coexistence in savanna. *J. Ecol.* **2000**, *88*, 213–229. [CrossRef]
4. Bond, W.J.; Woodward, F.I.; Midgley, G.F. The global distribution of ecosystems in a world without fire. *New Phytol.* **2005**, *165*, 525–538. [CrossRef] [PubMed]
5. Shorrocks, B.; Bates, W. *The Biology of African Savannahs*; Oxford University Press: New York, NY, USA, 2007.
6. Maurin, O.; Davies, T.J.; Burrows, J.E.; Daru, B.H.; Yessoufou, K.; Muasya, A.M.; van der Bank, M.; Bond, W.J. Savanna fire and the origins of the 'underground forests' of Africa. *New Phytol.* **2014**, *204*, 201–214. [CrossRef]
7. Ferreira, L.N.; Vega-Oliveros, D.A.; Zhao, L.; Cardoso, M.F.; Macau, E.E. Global fire season severity analysis and forecasting. *Comput. Geosci.* **2020**, *134*, 104339. [CrossRef]
8. Van der Werf, G.R.; Randerson, J.T.; Giglio, L.; Collatz, G.J.; Mu, M.; Kasibhatla, P.S.; Morton, D.C.; DeFries, R.S.; Jin, Y.; van Leeuwen, T.T. Global fire emissions and the contribution of deforestation, savanna, forest, agricultural, and peat fires (1997–2009). *Atmos. Chem. Phys.* **2010**, *10*, 11707–11735. [CrossRef]
9. Giglio, L.; Randerson, J.T.; van der Werf, G.R. Analysis of daily, monthly, and annual burned area using the fourth-generation global fire emissions database (GFED4). *J. Geophys. Res.* **2013**, *118*, 317–328. [CrossRef]
10. Grégoire, J.M.; Tansey, K.; Silva, J.M.N. The GBA2000 initiative: Developing a global burnt area database from SPOT-VEGETATION imagery. *Int. J. Remote Sens.* **2003**, *24*, 1369–1376. [CrossRef]
11. Goldammer, J.G.; De Ronde, C. *Wildland fire Management Handbook for Sub-Sahara Africa*; Global Fire Monitoring Center: Freiburg, Germany, 2004.
12. Hodnebrog, Ø.; Myhre, G.; Forster, P.M.; Sillmann, J.; Samset, B.H. Local biomass burning is a dominant cause of the observed precipitation reduction in southern Africa. *Nat. Commun.* **2016**, *7*, 1–8. [CrossRef]
13. Schneibel, A.; Stellmes, M.; Röder, A.; Finckh, M.; Revermann, R.; Frantz, D.; Hill, J. Evaluating the trade-off between food and timber resulting from the conversion of Miombo forests to agricultural land in Angola using multi-temporal Landsat data. *Sci. Total Environ.* **2016**, *548*, 390–401. [CrossRef] [PubMed]
14. United States Agency for International Development. Angola Biodiversity and Tropical Forests: 118/119 Assessment. Prepared by the Biodiversity Analysis and Technical Support (BATS) Team for the United States Agency for International Development. Available online: https://usaidgems.org/Documents/FAA&Regs/FAA118119/Angola2013.pdf (accessed on 6 April 2020).
15. Archibald, S.; Roy, D.P.; van Wilgen, B.W.; Scholes, R.J. What limits fire? An examination of drivers of burnt area in Southern Africa. *Glob. Change Biol.* **2009**, *15*, 613–630. [CrossRef]
16. Shlisky, A.; Alencar, A.A.; Nolasco, M.M.; Curran, L.M. Overview: Global fire regime conditions, threats, and opportunities for fire management in the tropics. In *Tropical Fire Ecology*; Cochrane, M.A., Ed.; Springer–Praxis: Berlin, Germany, 2009; pp. 65–83.
17. Barbosa, P.M.; Stroppiana, D.; Grégoire, J.M.; Pereira, J.M.C. An assessment of vegetation fire in Africa (1981–1991): Burned areas, burned biomass, and atmospheric emissions. *Global Biogeochem. Cycles* **1999**, *13*, 933–950. [CrossRef]
18. Archibald, S.; Scholes, R.J.; Roy, D.P.; Roberts, G.; Boschetti, L. Southern African fire regimes as revealed by remote sensing. *Int. J. Wildland Fire* **2010**, *19*, 861–878. [CrossRef]
19. Frazão, R.; Catarino, S.; Goyder, D.; Darbyshire, I.; Magalhães, M.F.; Romeiras, M.M. Species richness and distribution of the largest plant radiation of Angola: *Euphorbia* (Euphorbiaceae). *Biodivers. Conserv.* **2020**, *29*, 187–206. [CrossRef]
20. Olson, D.M.; Dinerstein, E.; Wikramanayake, E.D.; Burgess, N.D.; Powell, G.V.N.; Underwood, E.C.; D'amico, J.A.; Itoua, I.; Strand, H.E.; Morrison, J.C.; et al. Terrestrial ecoregions of the world a new map of life on earth a new global map of terrestrial ecoregions provides an innovative tool for conserving biodiversity. *Bioscience* **2001**, *51*, 933–938. [CrossRef]
21. Romeiras, M.M.; Figueira, R.; Duarte, M.C.; Beja, P.; Darbyshire, I. Documenting biogeographical patterns of African timber species using herbarium records: A conservation perspective based on native trees from Angola. *PLoS ONE* **2014**, *9*, e103403. [CrossRef]

22. Dwyer, E.; Pinnock, S.; Grégoire, J.M.; Pereira, J.M.C. Global spatial and temporal distribution of vegetation fire as determined from satellite observations. *Int. J. Remote Sens.* **2020**, *21*, 1289–1302. [CrossRef]

23. Mendelsohn, J.M. Landscape changes in Angola. In *Biodiversity of Angola. Science & Conservation: A Modern Synthesis*; Huntley, B.J., Russo, V., Lages, F., Ferrand, N., Eds.; Springer: Cham, Switzerland, 2019; pp. 123–137. [CrossRef]

24. Mishra, N.B.; Mainali, K.P.; Crews, K.A. Modelling spatiotemporal variability in fires in semiarid savannas: A satellite-based assessment around Africa's largest protected area. *Int. J. Wildland Fire* **2016**, *25*, 730–741. [CrossRef]

25. Schneibel, A.; Frantz, D.; Röder, A.; Stellmes, M.; Fischer, K.; Hill, J. Using annual landsat time series for the detection of dry forest degradation processes in south-central Angola. *Remote Sens.* **2017**, *9*, 905. [CrossRef]

26. De Cauwer, V.; Mertens, J. Impact of fire on the Baikiaea woodlands. In *Climate Change and Adaptive Land Management in Southern Africa—Assessments, Changes, Challenges, and Solutions*; Revermann, R., Krewenka, K.M., Schmiedel, U., Olwoch, J.M., Helmschrot, J., Jürgens, N., Eds.; Biodiversity & Ecology, Klaus Hess Publishers: Windhoek, Namibia, 2018; pp. 334–335. [CrossRef]

27. Sankaran, M.; Ratnam, J.; Hanan, N.P. Tree-grass coexistence in savannas revisited—Insights from an examination of assumptions and mechanisms invoked in existing models. *Ecol. Lett.* **2004**, *7*, 480–490. [CrossRef]

28. Gignoux, J.; Lahoreau, G.; Julliard, R.; Barot, S. Establishment and early persistence of tree seedlings in an annually burned savanna. *J. Ecol.* **2009**, *97*, 484–495. [CrossRef]

29. Stellmes, M.; Frantz, D.; Finckh, M.; Revermann, R. Fire frequency, fire seasonality and fire intensity within the Okavango region derived from MODIS fire products. *Biodivers. Ecol.* **2013**, *5*, 351–362. [CrossRef]

30. Hansen, M.C.; Potapov, P.V.; Moore, R.; Hancher, M.; Turubanova, S.A.; Tyukavina, A.; Thau, D.; Stehman, S.V.; Goetz, S.J.; Loveland, T.R.; et al. High-Resolution Global Maps of 21st-Century Forest Cover Change. *Science* **2013**, *342*, 850–853. [CrossRef]

31. Food and Agriculture Organization of the United Nations. Global Forest Resources Assessment 2020—Key Findings. Available online: http://www.fao.org/3/CA8753EN/CA8753EN.pdf (accessed on 8 June 2020).

32. Global Forest Watch. Available online: https://www.globalforestwatch.org/map (accessed on 3 March 2020).

33. Graham, P.H.; Vance, C.P. Legumes: Importance and constraints to greater use. *Plant Physiol.* **2003**, *131*, 872–877. [CrossRef] [PubMed]

34. Catarino, S.; Duarte, M.C.; Costa, E.; Carrero, P.G.; Romeiras, M.M. Conservation and sustainable use of the medicinal Leguminosae plants from Angola. *PeerJ* **2019**, *7*, e6736. [CrossRef]

35. Syampungani, S.; Chirwa, P.W.; Akinnifesi, F.K.; Sileshi, G.; Ajayi, O.C. The miombo woodlands at the cross roads: Potential threats, sustainable livelihoods, policy gaps and challenges. *Nat. Resour. Forum.* **2009**, *33*, 150–159. [CrossRef]

36. Ribeiro, N.S.; Syampungani, S.; Nangoma, D.; Ribeiro-Barros, A. Miombo Woodlands Research Towards the Sustainable use of Ecosystem Services in Southern Africa. In *Biodiversity in Ecosystems-Linking Structure and Function*; Lo, Y., Blanco, J.A., Roy, S., Eds.; IntechOpen: London, UK, 2015; pp. 475–491.

37. Jew, E.K.; Dougill, A.J.; Sallu, S.M.; O'Connell, J.; Benton, T.G. Miombo woodland under threat: Consequences for tree diversity and carbon storage. *For. Ecol. Manag.* **2016**, *361*, 144–153. [CrossRef]

38. Climate Change 2014: Impacts, Adaptation, and Vulnerability. Part B: Regional Aspects. Fifth Assessment Report of the Intergovernmental Panel on Climate Change. Available online: https://www.ipcc.ch/site/assets/uploads/2018/02/WGIIAR5-PartB_FINAL.pdf (accessed on 3 April 2020).

39. Enright, N.J.; Fontaine, J.B.; Bowman, D.M.; Bradstock, R.A.; Williams, R.J. Interval squeeze: Altered fire regimes and demographic responses interact to threaten woody species persistence as climate changes. *Front. Ecol. Environ.* **2015**, *13*, 265–272. [CrossRef]

40. Shlisky, A.; Waugh, J.; Gonzalez, P.; Gonzalez, M.; Manta, M.; Santoso, H.; Alvarado, E.; Nuruddin, A.A.; Rodriguez-Trejo, D.A.; Swarty, R.; et al. *Fire, Ecosystems and People: Threats and Strategies for Global Biodiversity Conservation. Global Fire Initiative Technical Report 2007-2*; The Nature Conservancy: Arlington, VA, USA, 2007.

41. Silva, J.M.N.; Pereira, J.M.C.; Cabral, A.I.; Sá, A.C.L.; Vasconcelos, M.J.P.; Mota, B.; Grégoire, J.M. An estimate of the area burned in southern Africa during the 2000 dry season using SPOT-VEGETATION satellite data. *J. Geophys. Res.* **2003**, *108*, 8498. [CrossRef]

42. Grégoire, J.M.; Eva, H.D.; Belward, A.S.; Palumbo, I.; Simonetti, D.; Brink, A. Effect of land-cover change on Africa's burnt area. *Int. J. Wildland Fire* **2013**, *22*, 107–120. [CrossRef]

43. Giglio, L.; van der Werf, G.R.; Randerson, J.T.; Collatz, G.J.; Kasibhatla, P. Global estimation of burned area using MODIS active fire observations. *Atmos. Chem. Phys.* **2006**, *6*, 957–974. [CrossRef]

44. Giglio, L.; Boschetti, L.; Roy, D.P.; Humber, M.L.; Justice, C.O. The Collection 6 MODIS burnt area mapping algorithm and product. *Remote Sens. Environ.* **2018**, *217*, 72–85. [CrossRef]

45. Roy, D.P.; Boschetti, L. Southern Africa Validation of the MODIS, L3JRC, and GlobCarbon Burned-Area Products. *Geosci. Remote* **2009**, *47*, 1032–1044. [CrossRef]

46. Huntley, B.J. Angola in Outline: Physiography, Climate and Patterns of Biodiversity. In *Biodiversity of Angola Science & Conservation: A Modern Synthesis*; Huntley, B.J., Russo, V., Lages, F., Ferrand, N., Eds.; Springer: Cham, Switzerland, 2019; pp. 15–42. [CrossRef]

47. Ministério do Ambiente de Angola. *5° Relatório Nacional 2007–2012 Sobre a Implementação da Convenção da Diversidade Biológica em Angola*; Direcção Nacional da Biodiversidade, República de Angola: Luanda, Angola, 2014.

48. Huntley, B.J.; Beja, P.; Vaz-Pinto, P.; Russo, V.; Veríssimo, L.; Morais, M. Biodiversity conservation history, protected areas and hotspots. In *Biodiversity of Angola. Science & Conservation: A Modern Synthesis*; Huntley, B.J., Russo, V., Lages, F., Ferrand, N., Eds.; Springer: Cham, Switzerland, 2019; pp. 495–512. [CrossRef]

49. Instituto Nacional de Estatística. *Resultados Preliminares do Recenseamento Geral da População e da Habitação de Angola 2014*; Instituto Nacional de Estatística: Luanda, Angola, 2014.

50. Pröpper, M.; Gröngröft, A.; Finckh, M.; Stirn, S.; De Cauwer, V. *The Future Okavango: Findings, Scenarios and Recommendations for Action: Research Project Final Synthesis Report 2010–2015*; University of Hamburg-Biocentre Klein Flottbek: Hamburg, Germany, 2015.

51. Goyder, D.J.; Gonçalves, F.M.P. The Flora of Angola: Collectors, richness and endemism. In *Biodiversity of Angola. Science & Conservation: A Modern Synthesis*; Huntley, B.J., Russo, V., Lages, F., Ferrand, N., Eds.; Springer: Cham, Switzerland, 2019; pp. 79–96. [CrossRef]

52. MCD64A1 MODIS/Terra+Aqua Burnt Area Monthly L3 Global 500m SIN Grid V006 [Data set]. NASA EOSDIS Land Processes DAAC. Available online: https://lpdaac.usgs.gov/products/mcd64a1v006/ (accessed on 3 February 2020). [CrossRef]

53. Vermote, E.F.; Justice, N.Z.E. Operational atmospheric correction of the MODIS data in the visible to middle infrared: First results. *Remote Sens. Environ.* **2002**, *83*, 97–111. [CrossRef]

54. Justice, C.O.; Giglio, L.; Korontzi, S.; Owens, J.; Morisette, J.T.; Roy, D.; Descloitres, J.; Alleaume, S.; Petitcolin, F.; Kaufman, Y. The MODISfire products. *Remote Sens. Environ.* **2002**, *83*, 244–262. [CrossRef]

55. Xiong, X.; Chiang, K.; Sun, J.; Barnes, W.L.; Guenther, B.; Salomonson, V.V. NASA EOS Terra and Aqua MODIS on-orbit performance. *Adv. Space Res.* **2009**, *43*, 413–422. [CrossRef]

56. Roy, D.P.; Jin, Y.; Lewis, P.E.; Justice, C.O. Prototyping a global algorithm for systematic fire affected area mapping using MODIS time series data. *Remote Sens. Environ.* **2005**, *97*, 137–162. [CrossRef]

57. QGIS Development Team. QGIS Geographic Information System. Open Source Geospatial Foundation Project. Available online: http://qgis.osgeo.org (accessed on 24 February 2020).

58. TerrSet 2020, Geospatial Monitoring and Modeling System. Clark Labs. Available online: https://clarklabs. org/wp-content/uploads/2020/05/TerrSet_2020_Brochure-FINAL27163334.pdf (accessed on 5 February 2020).

59. Durbin, J.; Watson, G.S. Testing for serial correlation in least squares regression. *Biometrika* **1950**, *37*, 409–428. [CrossRef]

60. Yue, S.; Wang, C.Y. Regional streamflow trend detection with consideration of both temporal and spatial correlation. *Int. J. Climatol.* **2002**, *22*, 933–946. [CrossRef]

61. Silva, J.M.N.; Moreno, M.V.; Le Page, Y.; Oom, D.; Bistinas, I.; Pereira, J.M.C. Spatiotemporal trends of area burnt in the Iberian Peninsula, 1975–2013. *Reg. Environ. Change* **2019**, *19*, 515–527. [CrossRef]

62. Wang, X.L.L.; Swail, V.R. Changes of extreme wave heights in Northern Hemisphere oceans and related atmospheric circulation regimes. *J. Clim.* **2001**, *14*, 2204–2221. [CrossRef]

63. Mann, H.B. Nonparametric tests against trend. *Econometrica* **1945**, *13*, 245–259. [CrossRef]

64. Kendall, M.G. *Rank Correlation Methods*, 2nd ed.; C. Griffin: London, UK, 1975.

65. Neeti, N.; Eastman, J.R. A contextual Mann-Kendall approach for the assessment of trend significance in image time series. *Trans. GIS* **2011**, *15*, 599–611. [CrossRef]

66. Chandler, R.E.; Scott, E.M. *Statistical Methods for Trend Detection and Analysis in the Environmental Sciences*; Wiley: Chichester, UK, 2011.

67. Theil, H. A rank-invariant method of linear and polynomial regression analysis. I, II, III. *Proc. R. Neth. Acad. Arts Sci.* **1950**, *53*, 386–392, 521–525, 1397–1412. [CrossRef]

68. Sen, P.K. Estimates of the regression coefficient based on Kendall's tau. *J. Am. Stat. Assoc.* **1968**, *63*, 1379–1389. [CrossRef]

69. Land Cover CCI: Product User Guide Version 2.0. Available online: http://maps.elie.ucl.ac.be/CCI/viewer/download/ESACCI-LC-Ph2-PUGv2_2.0.pdf (accessed on 2 March 2020).

70. Defourny, P.; Lamarche, C.; Flasse, C. *Product User Guide and Specification ICDR Land Cover 2016 and 2017. Version 1.1.1.*; ECMWF: Shinfield Park, UK, 2019.

71. Terrestrial Ecoregions of the World. Available online: https://www.worldwildlife.org/publications/terrestrial-ecoregions-of-the-world (accessed on 3 March 2020).

72. Burgess, N.; Hales, J.A.; Underwood, E.; Dinerstein, E.; Olson, D.; Itoua, I.; Schipper, J.; Rickketts, T.; Newman, K. *Terrestrial Ecoregions of Africa and Madagascar: A Conservation Assessment*; Island Press: Washington, DC, USA, 2004.

73. Ladle, R.J.; Whittaker, R.J. *Conservation Biogeography*; Wiley-Blackwell: Oxford, UK, 2011.

74. Rodrigues, P.; Figueira, R.; Vaz Pinto, P.; Araújo, M.B.; Beja, P. A biogeographical regionalization of Angolan mammals. *Mamm. Rev.* **2015**, *45*, 103–116. [CrossRef]

75. World Database on Protected Areas. Available online: https://www.protectedplanet.net/c/world-database-on-protected-areas (accessed on 3 March 2020).

76. Land Cover Classification Gridded Maps from 1992 to Present Derived from Satellite Observations. Available online: https://cds.climate.copernicus.eu/cdsapp#!/dataset/satellite-land-cover?tab=overview (accessed on 3 March 2020).

77. Andela, N.; van Der Werf, G.R. Recent trends in African fires driven by cropland expansion and El Nino to La Nina transition. *Nat. Clim. Chang.* **2014**, *4*, 791–795. [CrossRef]

78. Andela, N.; van der Werf, G.R.; Kaiser, J.W.T.; van Leeuwen, T.; Wooster, M.J.; Lehmann, C.E.R. Biomass burning fuel consumption dynamics in the tropics and subtropics assessed from satellite. *Biogeosciences* **2016**, *13*, 3717–3734. [CrossRef]

79. Myers, N.; Mittermeier, R.A.; Mittermeier, C.G.; da Fonseca, G.A.B.; Kent, J. Biodiversity hotspots for conservation priorities. *Nature* **2000**, *403*, 853–858. [CrossRef]

80. Carranza, F.; Treakle, J.; Groppo, P. *Land, Territorial Development and Family Farming in Angola: A Holistic Approach to Community-based Natural Resource Governanance: The Cases of Bie, Huambo, and Huila Provinces*; Food and Agriculture Organization of the United Nations: Rome, Italy, 2014.

81. Huntley, B.J. *Wildlife at War in Angola. The Rise and Fall of an African Eden*; Protea Book House: Pretoria, South Africa, 2017.

82. Peel, M.C.; Finlayson, B.L.; Mcmahon, T.A. Updated world map of the Köppen Geiger climate classification. *Hydrol. Earth Syst. Sci.* **2007**, *11*, 1633–1644. [CrossRef]

83. Mills, M.S.; Dean, W.R.J.; Town, R. The avifauna of the Lagoa Carumbo area, northeast Angola. *Malimbus* **2013**, *35*, 77–92.

84. Maquia, I.; Catarino, S.; Pena, A.R.; Brito, D.R.; Ribeiro, N.; Romeiras, M.M.; Ribeiro-Barros, A.I. Diversification of African Tree Legumes in Miombo–Mopane Woodlands. *Plants* **2019**, *8*, 182. [CrossRef] [PubMed]

85. Timberlake, J.; Chidumayo, E. *Miombo Ecoregion Vision Report. Occasional Publications in Biodiversity no. 20*; Biodiversity Foundation for Africa: Bulawayo, Zimbabwe, 2011.

86. White, F. *The Vegetation of Africa*; Unesco: Paris, France, 1983.

87. Palacios, G.; Lara-Gomez, M.; Márquez, A.; Vaca, J.L.; Ariza, D.; Lacerda, V.; Navarro-Cerrillo, R.M. *Miombo's Cover Change in Huambo Province (2002–2015)*; SASSCAL Project Proceedinds: Huambo, Angola, 2015.

88. Schneibel, A.; Stellmes, M.; Revermann, R.; Finckh, M. Agricultural expansion during the post-civil war period in southern Angola based on bi-temporal Landsat data. *Biodivers. Ecol.* **2013**, *5*, 311–320. [CrossRef]

89. Mendelsohn, J.M.; Mendelsohn, S. *Sudoeste de Angola: Um Retrato da Terra e da Vida. South West Angola: A Portrait of Land and Life*; Raison: Windhoek, Namibia, 2018.

90. Food and Agriculture Organization of the United Nations. Global Forest Resources Assessment 2010, Main Report. Available online: http://www.fao.org/3/a-i1757e.pdf (accessed on 8 June 2020).

91. Li, W.; Ciais, P.; MacBean, N.; Peng, S.; Defourny, P.; Bontemps, S. Major forest changes and land cover transitions based on plant functional types derived from the ESA CCI Land Cover product. *Int. J. Appl. Earth Obs. Geoinf.* **2016**, *47*, 30–39. [CrossRef]

92. Li, W.; MacBean, N.; Ciais, P.; Defourny, P.; Lamarche, C.; Bontemps, S.; Houghton, R.A.; Peng, S. Gross and net land cover changes in the main plant functional types derived from the annual ESA CCI land cover maps (1992–2015). *Earth Syst. Sci. Data* **2018**, *10*, 219–234. [CrossRef]

93. Gonçalves, F.M.; Revermann, R.; Gomes, A.L.; Aidar, M.P.; Finckh, M.; Juergens, N. Tree species diversity and composition of miombo woodlands in south-central Angola: A chronosequence of forest recovery after shifting cultivation. *Int. J. For. Res.* **2017**, *2017*, 6202093. [CrossRef]

94. Chase, M.J.; Griffin, C.R. Elephants of south-east Angola in war and peace: Their decline, re-colonization and recent status. *Afr. J. Ecol.* **2011**, *49*, 353–361. [CrossRef]

95. Eby, S.L.; Dempewolf, J.; Holdo, R.M.; Metzger, K.L. Fire in the Serengeti ecosystem: History, drivers, and consequences. In *Serengeti IV: Sustaining Biodiversity in a Coupled Human-Natural System*; de Sinclair, A.R.E., Metzger, K.L., Mduma, S.A.R., Fryxell, J.M., Eds.; The University of Chicago Press: Chicago, IL, USA, 2015.

96. Butler, B.O.; Ceríaco, L.M.; Marques, M.P.; Bandeira, S.; Júlio, T.; Heinicke, M.; Bauer, A.M. Herpetological survey of Huíla Province, Southwest Angola, including first records from Bicuar National Park. *Herpetol. Rev.* **2019**, *50*, 225–240.

97. Pinto, P.V. The Giant Sable Antelope: Angola's National Icon. In *Biodiversity of Angola. Science & Conservation: A Modern Synthesis*; Huntley, B.J., Russo, V., Lages, F., Ferrand, N., Eds.; Springer: Cham, Switzerland, 2019; pp. 471–491. [CrossRef]

98. Ministério do Urbanismo e Ambiente. *National Biodiversity Strategy and Action Plan (2007–2012)*; Ministério do Urbanismo e Ambiente, República de Angola: Luanda, Angola, 2006.

99. Austen, B. The History of Veld Burning in the Wankie National Park, Rhodesia. *Proc. Annu. Tall Timbers Fire Ecol. Conf.* **1972**, *11*, 277–296.

100. Venter, F.J.; Naiman, R.J.; Biggs, H.C.; Pienaar, D.J. The evolution of conservation management philosophy: Science, environmental change and social adjustments in Kruger National Park. *Ecosystems* **2008**, *11*, 173–192. [CrossRef]

101. te Beest, M.; Cromsigt, J.P.G.M.; Ngobese, J.; Olff, H. Managing invasions at the cost of native habitat? An experimental test of the impact of fire on the invasion of Chromolaena odorata in a South African savanna. *Biol. Invasions* **2012**, *14*, 607–618. [CrossRef]

MDPI

St. Alban-Anlage 66

4052 Basel

Switzerland

Tel. +41 61 683 77 34

Fax +41 61 302 89 18

www.mdpi.com

Diversity Editorial Office

E-mail: diversity@mdpi.com

www.mdpi.com/journal/diversity